MATHEMATICAL MODELS FOR SYSTEMS RELIABILITY

MATHEMATICAL MODELS FOR SYSTEMS RELIABILITY

Benjamin Epstein
Ishay Weissman

CRC Press
Taylor & Francis Group
Boca Raton London New York

CRC Press is an imprint of the
Taylor & Francis Group, an **informa** business
A CHAPMAN & HALL BOOK

Chapman & Hall/CRC
Taylor & Francis Group
6000 Broken Sound Parkway NW, Suite 300
Boca Raton, FL 33487-2742

First issued in paperback 2019

ISBN-13: 978-1-4200-8082-7 (hbk)
ISBN-13: 978-0-367-38732-7 (pbk)

Library of Congress Cataloging-in-Publication Data

Epstein, Benjamin, 1918-
    Mathematical models for systems reliability / Benjamin Epstein and Ishay Weissman.
        p. cm.
    "A CRC title."
    Includes bibliographical references and index.
    ISBN 978-1-4200-8082-7 (alk. paper)
    1. Reliability (Engineering)--Mathematics. 2. System failures (Engineering)--Mathematical models. I. Weissman, Ishay, 1940- II. Title.

TA169.E67 2008
620'.00452015118--dc22   2008010874

Visit the Taylor & Francis Web site at
http://www.taylorandfrancis.com

and the CRC Press Web site at
http://www.crcpress.com

Dedicated to

my wife **Edna**

and my children **Shoham, Tsachy** and **Rom**

Ishay Weissman

Preface

This book has evolved from the lectures of Professor Benjamin (Ben) Epstein (1918–2004) at the Technion—Israel Institute of Technology. Throughout his tenure at the Technion, from 1968 until his retirement in 1986, he designed and taught two courses on Reliability Theory. One, which he considered to be fundamental in reliability considerations, was *Mathematical Models for Systems Reliability*. The second course was *Statistical Methods in Reliability*. As these titles indicate, although there was some overlapping, the first course concentrated on the mathematical probabilistic models while the second course concentrated on statistical data analysis and inference applied to systems reliability.

Epstein was one of the pioneers in developing the theory of reliability. He was the first to advocate the use of the exponential distribution in life-testing and developed the relevant statistical methodology. Later on, when Sobel Milton joined Epstein's Math Department in Wayne State, they published their joint work in a sequence of papers. Here is what Barlow and Proschan say in their now classical book [3]:

> In 1951 Epstein and Sobel began work in the field of life-testing which was to result in a long stream of important and extremely influential papers. This work marked the beginning of the widespread assumption of the exponential distribution in life-testing research.

Epstein's contributions were officially recognized in 1974, when the American Society for Quality decorated him with the prestigious Shewhart Medal.

Epstein's lecture notes for *Mathematical Models for Systems Reliability* have never been published. However, in 1969 they were typed, duplicated and sold to Technion students by the Technion Student Association. Soon enough, they were out of print, but luckily, five copies remained in the library, so students could still use (or copy) them. After Epstein's retirement and over the last two decades, I taught the course, using Epstein's notes. During the years, I added some more topics, examples and problems, gave alternative proofs to some results, but the general framework remained Epstein's. In view of the fact that the *Statistical Methods in Reliability* course was no longer offered, I added a brief introduction to Statistical Estimation Theory, so that the students could relate to estimation aspects in reliability problems (mainly in Chapters 1–3 and 6).

It is my conviction, that the material presented in this book provides a rigorous treatment of the required probability background for understanding reliability theory. There are many contemporary texts available in the market, which emphasize other aspects of reliability, as statistical methods, life-testing, engineering, reliability of electronic devices, mechanical devices, software reliability, etc. The interested reader is advised to Google the proper keywords to find the relevant literature.

The book can serve as a text for a one-semester course. It is assumed that the readers of the book have taken courses in Calculus, Linear Algebra and Probability Theory. Knowledge of Statistical Estimation, Differential Equations and Laplace Transform Methods are advantageous, though not necessary, since the basic facts needed are included in the book.

The Poisson process and its associated probability laws are important in reliability considerations and so it is only natural that Chapter 1 is devoted to this topic. In Chapter 2, a number of stochastic models are considered as a framework for discussing life length distributions. The fundamental concept of the hazard or force of mortality function is also introduced in this chapter. Formal rules for computing the reliability of non-repairable systems possessing commonly occurring structures are given in Chapter 3. In Chapter 4 we discuss the stochastic behavior over time of one-unit repairable systems and such measures of system effectiveness as point-availability, interval and long-run availability and interval reliability are introduced. The considerations of Chapter 4 are extended to two-unit repairable systems in Chapter 5. In Chapter 6 we introduce the general continuous-time Markov chains, pure birth and death processes and apply the results to n-unit repairable systems. We introduce the transitions and rates diagrams and present several methods for computing the transition probabilities matrix, including the use of computer software. First passage time problems are considered in Chapter 7 in the context of systems reliability. In Chapters 8 and 9 we show how techniques involving the use of embedded Markov chains, semi-Markov processes, renewal processes, points of regeneration and integral equations can be applied to a variety of reliability problems, including preventive maintenance.

I am extremely grateful to Malka Epstein for her constant encouragement and unflagging faith in this project. I would also like to thank the Technion at large, and the Faculty of Industrial Engineering and Management in particular, for having provided such a supportive and intellectually stimulating environment over the last three decades. Lillian Bluestein did a a superb job in typing the book, remaining constantly cheerful throughout.

Finally, and most importantly, I want to thank my wife and children for their continual love and support, without which I doubt that I would have found the strength to complete the book.

Ishay Weissman
Haifa, Israel
April 2008

Contents

CHAPTER 1

Preliminaries

1.1 The Poisson process and distribution

The Poisson process and its generalizations play a fundamental role in the kinds of reliability models that we shall be considering. In discussing what we mean by a Poisson process, it is very helpful to think of random phenomena such as the emission of α-particles by a radioactive substance, telephone calls coming into an exchange, customers arriving for service, machine breakdowns over time, etc. To be specific, let us imagine that we start observing some radioactive material with a Geiger counter at time $t = 0$, and record each emission as it occurs (we assume an idealized counter capable of recording each emission). Of special interest is $N(t)$, the number of emissions recorded on or before time t. Any particular realization of $N(t)$ is clearly a step function with unit jumps occurring at the times when an emission occurs. Thus, if emissions occur at $0 < t_1 < t_2 < t_3 < \cdots < t_k \ldots$, the associated $N(t)$ is

$$N(t) = 0, \qquad 0 \le t < t_1,$$
$$N(t) = 1, \qquad t_1 \le t < t_2,$$
$$N(t) = 2, \qquad t_2 \le t < t_3,$$

$$\vdots$$

$$N(t) = k, \qquad t_k \le t < t_{k+1},$$

etc.

Each act of observing the counter for a length of time t_0 produces a possible realization of the random function $N(t)$ in a time interval of length t_0. One such realization is shown in Figure 1.1. $N(t_0)$, the height of the random function $N(t)$ at a fixed time t_0 is a discrete random variable, which can take on the values $0, 1, 2, \ldots$. It is of interest to find the probability that $N(t_0) = n$, which we denote as $P[N(t_0) = n]$ or more succinctly as $P_n(t_0)$, for $n = 0, 1, 2, \ldots$.

In order to compute $P_n(t_0)$, one must make some assumptions about the way

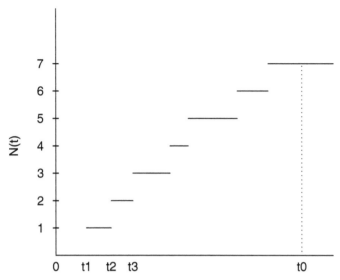

Figure 1.1 *A particular realization of $N(t)$ $(0 \le t \le t_0)$, the number of emissions up to time t_0.*

in which emissions occur over time. The simplest set of assumptions, which seems to be reasonable not only for emissions by a radioactive substance but in many other contexts, is the following:

(i) The number of emissions observed in two (or more) nonoverlapping time intervals are mutually independent random variables.

(ii) The probability of an emission in the time interval $(t, t + h]$, where $h > 0$ is small, is "approximately" λh, or more precisely $\lambda h + o(h)$. Here λ can be thought of physically as the emission rate per unit time.

(iii) The probability of two or more emissions in the time interval of length h is $o(h)$.

The term $o(h)$ in (ii) and (iii) means a quantity which goes to zero faster than h, namely, $o(h)/h \to 0$ as $h \to 0$.

A random process developing over time and meeting these three conditions (where the word *emission* would be replaced by whatever is appropriate) is said to be a *(temporally) homogeneous Poisson process*. The term *homogeneous* refers to our assuming in (ii) that λ is the same for all time. Assumptions (i) and (ii) taken together state that the probability of an emission in $(t, t + h]$ is approximately λh independent of how many emissions have occurred in $(0, t]$. Assumption (iii) states that, for small h, the probability of observing a clustering of two or more emissions in $(t, t+h]$ is "negligible" when compared with the probability of occurrence of a single emission.

The reader is reminded that a discrete random variable X, which takes on the integer values $0, 1, 2, \ldots, k, \ldots$ with probabilities

$$P[X = k] = e^{-\alpha} \frac{\alpha^k}{k!} \quad (\alpha > 0),$$

is said to be a *Poisson random variable* with parameter α. We now prove that, for the homogeneous Poisson process, for each fixed $t > 0$, $N(t)$ is a Poisson random variable with parameter λt, i.e.,

$$P_n(t) = P[N(t) = n] = e^{-\lambda t} \frac{(\lambda t)^n}{n!}, \quad n = 0, 1, 2, \ldots. \quad (1.1)$$

To prove Equation (1.1) let us consider two adjacent intervals $(0, t]$ and $(t, t + h]$. We first show how we can relate $P_0(t + h)$ to $P_0(t)$. To do this we note that the event $\{N(t + h) = 0\}$ (no emissions in $(0, t + h]$) takes place if and only if the events $\{N(t) = 0\}$ (no emission in $(0, t]$) and $\{N(t + h) - N(t) = 0\}$ (no emission in $(t, t + h]$) both occur. Hence, it follows from the independence assumption (i) that

$$P[N(t + h) = 0] = P[N(t) = 0] \cdot P[N(t + h) - N(t) = 0]. \quad (1.2)$$

It is an immediate consequence of assumption (ii) that

$$P[N(t + h) - N(t) = 0] = 1 - \lambda h + o(h). \quad (1.3)$$

Combining Equations (1.2) and (1.3) we get

$$P_0(t + h) = P_0(t)(1 - \lambda h) + o(h), \quad (1.4)$$

or

$$\frac{P_0(t + h) - P_0(t)}{h} = -\lambda P_0(t) + \frac{o(h)}{h}. \quad (1.5)$$

Letting $h \to 0$, we are led to the differential equation

$$P_0'(t) = -\lambda P_0(t). \quad (1.6)$$

Similarly we can relate $P_n(t + h)$ to $P_n(t)$ and $P_{n-1}(t)$ for $n \geq 1$ and by going to the limit as $h \to 0$ to obtain a differential equation expressing $P_n'(t)$ in terms of $P_n(t)$ and $P_{n-1}(t)$. For the case where $n = 1$ ($n \geq 2$), there are two (three) mutually exclusive ways of observing the event, $\{N(t + h) = n\}$ (exactly n emissions occur in $(0, t + h]$). These are:

(a) $\{N(t) = n\}$ and $\{N(t + h) - N(t) = 0\}$ occur
   or

(b) $\{N(t) = n - 1\}$ and $\{N(t + h) - N(t) = 1\}$ occur
   or

(c) $\{N(t) = n - k\}$ and $\{N(t + h) - N(t) = k\}$ occur, $\quad 2 \leq k \leq n$.
(If $n = 1$, only (a) or (b) can occur.)

It follows from the assumptions for a homogeneous Poisson process that the probabilities associated with (a), (b), and (c), respectively, are:

$$P_n(t)[1 - \lambda h + o(h)], \qquad P_{n-1}(t)[\lambda h + o(h)], \qquad o(h). \tag{1.7}$$

Hence, using the theorem of total probability, and combining terms of $o(h)$, we get

$$P_n(t + h) = P_n(t)[1 - \lambda h] + P_{n-1}(t)\lambda h + o(h) \tag{1.8}$$

or

$$\frac{P_n(t + h) - P_n(t)}{h} = -\lambda P_n(t) + \lambda P_{n-1}(t) + \frac{o(h)}{h}. \tag{1.9}$$

Letting $h \to 0$, Equation (1.9) becomes

$$P'_n(t) = -\lambda P_n(t) + \lambda P_{n-1}(t), \qquad n \geq 1. \tag{1.10}$$

We thus get the system of differential Equations (1.6) and (1.10) satisfied by $P_n(t)$. We also have the initial conditions $P_0(0) = 1$ and $P_n(0) = 0, n \geq 1$ (i.e., we assume that $N(t) = 0$ at $t = 0$).

The solution to Equation (1.6) subject to the initial condition $P_0(0) = 1$ is well known (can be easily verified by differentiation) and is given by

$$P_0(t) = e^{-\lambda t}. \tag{1.11}$$

Once we know $P_0(t)$, Equation (1.10), for $n = 1$, is equivalent to

$$\{e^{\lambda t} P_1(t)\}' = \lambda.$$

Integrating both sides gives

$$P_1(t) = e^{-\lambda t}(\lambda t + c)$$

for some constant c. Since $P_1(0) = 0$, we must have $c = 0$.

Proceeding inductively, we assume that

$$P_k(t) = e^{-\lambda t} \frac{(\lambda t)^k}{k!} \tag{1.12}$$

holds for $0 \leq k \leq n$ and we shall prove that it holds for $k = n + 1$. Under the induction assumption, Equation (1.10) is equivalent to

$$e^{\lambda t}(P'_{n+1}(t) + \lambda P_{n+1}(t)) = \{e^{\lambda t} P_{n+1}(t)\}' = \frac{\lambda(\lambda t)^n}{n!}. \tag{1.13}$$

Again, integrating both sides of Equation (1.13) gives

$$e^{\lambda t} P_{n+1}(t) = \frac{\lambda^{n+1}}{n!} \left(\frac{t^{n+1}}{n + 1} + c \right),$$

and the initial condition $P_{n+1}(0) = 0$ implies that $c = 0$. Thus, we have proven the validity of Equation (1.12) for $k = n + 1$ and consequently for all integers $k \geq 0$, i.e., $N(t)$ is a Poisson random variable with parameter λt.

It should be noted that $E[N(t)]$, the expected number of emissions in $(0, t]$ is given by

$$E[N(t)] = \sum_{n=0}^{\infty} n P_n(t) = \lambda t. \tag{1.14}$$

It can also be verified that $\text{Var}[N(t)]$, the variance of the number of emissions in $(0, t]$, is given by

$$\text{Var}[N(t)] = E[N^2(t)] - \{E[N(t)]\}^2 = \sum_{n=0}^{\infty} n^2 P_n(t) - \lambda^2 t^2 = \lambda t. \tag{1.15}$$

The *independence of increments* assumption (i) and the *homogeneity* properties imply that the laws of large numbers and the central limit theorem hold true for $N(t)$, as $t \to \infty$. The reason being that if $N(a, b) = N(b) - N(a)$ denotes the number of emissions in the time interval $(a, b]$, then for $n < t \leq n + 1$,

$$N(n) = \sum_{i=1}^{n} N(i - 1, i) \leq N(t) \leq N(n + 1). \tag{1.16}$$

Since $N(n)$ and $N(n + 1)$ are sums of independent and identically distributed random variables, all the classical asymptotic theory (as $n \to \infty$) holds true also for the (continuous-time) stochastic process N. Specifically, as $t \to \infty$, we have

$$N(t) \to \infty \ a.s.,$$

$$\frac{N(t)}{t} \to \lambda \ a.s.$$

and

$$\frac{N(t) - \lambda t}{(\lambda t)^{1/2}} \xrightarrow{d} \mathcal{N}(0, 1).$$

The notation $\xrightarrow{d} \mathcal{N}(0, 1)$ stands for convergence in distribution to a standard normal random variable.

In the context of reliability, the word "emission" would be replaced by "failure" and λ would be the failure rate per unit time (assuming, of course, that a temporally homogeneous Poisson process is a reasonable stochastic model for describing the occurrence of successive failures over time). For an elegant discussion of the Poisson process, the reader is advised to see Chapter 17 of Feller [26].

1.2 Waiting time distributions for a Poisson process

In Section 1.1, we obtained the distribution of $N(t)$, the number of occurrences (emissions, failures, etc.) when a *temporally* homogeneous Poisson process is observed for a fixed length of time t. It is of interest to ask the following related question: How long does it take to observe a fixed number of occurrences, say k? We denote the length of wait to the k'th occurrence by the random variable T_k. The cumulative distribution function (c.d.f.) of T_k is denoted by $F_k(t)$, where

$$F_k(t) = P(T_k \leq t). \tag{1.17}$$

To find $F_k(t)$ we note that the two events, $\{T_k \leq t\}$ and $\{N(t) \geq k\}$, are equivalent. Consequently,

$$F_k(t) = P(N(t) \geq k) = \sum_{j=k}^{\infty} P_j(t) = \sum_{j=k}^{\infty} e^{-\lambda t} \frac{(\lambda t)^j}{j!}$$

$$= 1 - \sum_{j=0}^{k-1} e^{-\lambda t} \frac{(\lambda t)^j}{j!}, \qquad t \geq 0. \tag{1.18}$$

$f_k(t)$, the probability density function (p.d.f.) of T_k obtained by differentiating $F_k(t)$, is

$$f_k(t) = \frac{\lambda (\lambda t)^{k-1} e^{-\lambda t}}{(k-1)!}, \qquad t \geq 0. \tag{1.19}$$

Expression (1.19) is known as a *gamma distribution* with parameters k and λ, denoted $\Gamma(k, \lambda)$. It is also frequently referred to as an Erlang distribution of order k, in honor of the Danish scientist who did pioneer work in stochastic processes, particularly as they relate to the operation of telephone exchanges (see the book by E. Brockmeyer et al. [8]).

When $k = 1$, Equations (1.18) and (1.19) become, respectively,

$$F_1(t) = 1 - e^{-\lambda t}, \qquad t \geq 0 \tag{1.20}$$

and

$$f_1(t) = \lambda e^{-\lambda t}, \qquad t \geq 0. \tag{1.21}$$

A random variable having a p.d.f. (1.21) is said to follow the exponential distribution with parameter $\lambda > 0$, denoted by $exp(\lambda)$. Thus, we see that T_1, the waiting time until the first occurrence, follows the exponential distribution. It should be mentioned parenthetically that by virtue of the assumptions for a homogeneous Poisson process, the p.d.f. of the waiting time between any two consecutive occurrences is also given by Equation (1.21).

It is interesting to note that $f_k(t)$ can be obtained directly. Clearly the event "time of the k'th occurrence lies in $(t, t + h]$" can happen in two mutually exclusive ways. These are:

(i) $(k-1)$ occurrences in $(0, t]$ and exactly one occurrence in $(t, t+h]$.

(ii) $(k-j)$ occurrences in $(0, t]$ and j occurrences in $(t, t+h]$, $2 \leq j \leq k$.

From the assumptions for a homogeneous Poisson process, it follows that the probabilities associated with (i) and (ii), respectively, are:

$$\frac{e^{-\lambda t}(\lambda t)^{k-1}}{(k-1)!} \lambda h + o(h) \quad \text{and} \quad o(h). \tag{1.22}$$

Hence, we can write

$$P(t < T_k \leq t+h) = \frac{e^{-\lambda t}(\lambda t)^{k-1}\lambda h}{(k-1)!} + o(h) \tag{1.23}$$

$$f_k(t) = \lim_{h \to 0} \frac{P(t < T_k \leq t+h)}{h} \tag{1.24}$$

and so dividing both sides of Equation (1.23) by h and letting $h \to 0$, we get

$$f_k(t) = \frac{\lambda(\lambda t)^{k-1} e^{-\lambda t}}{(k-1)!}, \qquad t \geq 0. \tag{1.25}$$

It is useful to note that the expectation and variance of T_k are given, respectively, by

$$E(T_k) = k/\lambda \quad \text{and} \quad \text{Var}(T_k) = k/\lambda^2.$$

This can be shown directly from the definition, i.e.,

$$E(T_k) = \int_0^\infty t f_k(t) dt \tag{1.26}$$

and

$$\text{Var}(T_k) = \int_0^\infty t^2 f_k(t) dt - [E(T_k)]^2 \tag{1.27}$$

or by noting that

$$T_k = T_1 + (T_2 - T_1) + \cdots + (T_k - T_{k-1}). \tag{1.28}$$

But it follows from the assumption of a homogeneous Poisson process, that the random variables $\{T_1, T_2 - T_1, T_3 - T_2, \ldots, T_k - T_{k-1}\}$ are mutually independent and each distributed with the same p.d.f. $f_1(t) = \lambda e^{-\lambda t}$, $t \geq 0$. Hence

$$E(T_k) = kE(T_1) = k/\lambda \quad \text{and} \quad \text{Var}(T_k) = k\text{Var}T_1 = k/\lambda^2. \tag{1.29}$$

The expected waiting time to the first occurrence $E(T_1)$, or more generally (because of the homogeneity property) the expected waiting time between successive occurrences, is the *mean time between failures* (MTBF) of reliability and life testing. The MTBF is often denoted by the symbol $\theta = 1/\lambda$.

1.3 Statistical estimation theory

In this section we introduce briefly some basic concepts and ideas in estimation theory. For a detailed development of the theory the reader may wish to consult such texts as Bickel and Doksum [5] or Lehmann and Casella [39].

1.3.1 Basic ingredients

Suppose we carry out an experiment whose outcome $\mathbf{X}$ is random, $\mathbf{X} \in \mathcal{X}$, where $\mathcal{X}$ is the *sample space*, i.e., the collection of all possible outcomes of our experiment. Here $\mathbf{X}$ could be $\{N(s) : 0 \le s \le t\}$, the number of failures of a machine, from time 0 to time s for all $0 \le s \le t$. Another typical situation is $\mathbf{X} = (X_1, X_2, \ldots, X_n)$, where the X_i are independent measurements of n objects, chosen at random from a certain population. We assume that the probability law of $\mathbf{X}$ is known up to an unknown parameter $\theta \in \Theta$. The *parameter space* Θ is known. In this section, θ stands for a general parameter (not necessarily an MTBF). Concerning $N(t)$, for instance, we may assume that its probability law is of a homogeneous Poisson process with an unknown parameter λ and $\Theta = \{\lambda : \lambda > 0\}$. For the second example we may assume that each X_i is normally distributed with mean μ and variance σ^2 (we write $X_i \sim N(\mu, \sigma^2)$). Here $\theta = (\mu, \sigma)$ and $\Theta = \{(\mu, \sigma) : -\infty < \mu < \infty; \sigma > 0\}$.

A *statistic* $S = S(\mathbf{X})$ is any function of $\mathbf{X}$ (but not of θ). Let $\tau = g(\theta) \in \mathbb{R}^1$ be a quantity of interest. Suppose we choose a statistic $\hat{\tau} = \hat{\tau}(\mathbf{X})$ to estimate τ. Then, we say that $\hat{\tau}$ is an *unbiased* estimator of τ if

$$E_\theta \hat{\tau} = \tau \quad (\forall\, \theta \in \Theta).$$

For the case $\mathcal{X} \subseteq \mathbb{R}^n$, let $f_\theta(\mathbf{x})$ stand for the p.d.f. of $\mathbf{X}$ if it is a continuous variable and the probability distribution if $\mathbf{X}$ is discrete. For the continuous case

$$E_\theta \hat{\tau} = \int_{\mathcal{X}} \hat{\tau}(\mathbf{x}) f_\theta(\mathbf{x}) d\mathbf{x}$$

and for the discrete case

$$E_\theta \hat{\tau} = \Sigma_{\mathbf{x} \in \mathcal{X}}\, \hat{\tau}(\mathbf{x}) f_\theta(\mathbf{x}).$$

The *bias* of an estimator is defined to be

$$b_\theta(\hat{\tau}) = E_\theta \hat{\tau} - \tau.$$

Thus, for an unbiased estimator the bias is 0. To measure the performance of an estimator we define its *mean squared error* (MSE) by

$$E_\theta(\hat{\tau} - \tau)^2 = \mathrm{Var}_\theta \hat{\tau} + b_\theta^2(\hat{\tau}).$$

A good estimator is one whose MSE is small for the entire range $\theta \in \Theta$.

1.3.2 Methods of estimation

In this subsection we discuss two methods: *maximum likelihood* and *method of moments*. In Subsection 1.3.5 we introduce the *Rao-Blackwell* method.

Maximum likelihood method. Suppose that our experiment ended up with an outcome $\mathbf{X} = \mathbf{x}$. The likelihood function $L(\theta)$ is then the probability of this event (or the p.d.f. evaluated at $\mathbf{x}$ in the continuous case) as a function of θ. The *maximum likelihood estimator* (MLE) of θ is a value $\hat{\theta} \in \Theta$ which maximizes $L(\theta)$. That is,

$$L(\hat{\theta}) = \max_{\theta \in \Theta} L(\theta).$$

In most cases of interest $\hat{\theta}$ is unique. If one wants to estimate $\tau = g(\theta)$, then the MLE of τ is simply $\hat{\tau} = g(\hat{\theta})$. Here are two examples.

Exponential sample. Suppose $\mathbf{X} = (X_1, X_2, \ldots, X_n)$, where the X_i are independent and exponentially distributed with p.d.f $f_\lambda(x) = \lambda e^{-\lambda x}$ $(x > 0)$ or in short $X_i \sim \exp(\lambda)$. The likelihood function, due to independence, is given by

$$L(\lambda) = \prod_{i=1}^{n} f_\lambda(X_i) = \lambda^n e^{-\lambda \Sigma X_i} \tag{1.30}$$

and clearly this is maximized at

$$\hat{\lambda} = \frac{n}{\Sigma X_i} = \frac{1}{\bar{X}}.$$

Here $\bar{X}$ is the sample mean. The MLE of $\mu = E_\lambda X_i = 1/\lambda$ is simply $\hat{\mu} = 1/\hat{\lambda} = \bar{X}$.

Poisson process. Suppose $\mathbf{X} = \{N(s) : 0 \leq s \leq t\}$ counts the number of failures of a certain machine up to time s for all $0 \leq s \leq t$ (the clock is stopped during repair-time). We assume that N is a homogeneous Poisson process with an unknown rate λ. Observing $\mathbf{X}$ is equivalent to observing the vector $(T_1, T_2, \ldots, T_{N(t)})$, where the T_i are the failure-times. Suppose that in our experiment $N(t) = k$ and $T_i = t_i$ $(0 < t_1 < t_2 < \cdots < t_k \leq t)$. Let $h > 0$ be very small (e.g., $h < \min_i(t_{i+1} - t_i)$). Then,

$$P\{N(t) = k, t_1 - h < T_1 \leq t_1, t_2 - h < T_2 \leq t_2, \ldots, t_k - h < T_k \leq t_k\} \tag{1.31}$$

$$= P\{N(0, t_1 - h) = 0, N(t_1 - h, t_1) = 1, N(t_1, t_2 - h) = 0,$$
$$N(t_2 - h, t_2) = 1, \ldots, N(t_k - h, t_k) = 1, N(t_k, t) = 0\}$$

$$= e^{-\lambda(t_1 - h)} \cdot \lambda h e^{-\lambda h} \cdot e^{-\lambda(t_2 - h - t_1)} \cdot \lambda h e^{-\lambda h} \cdot \ldots \cdot e^{-\lambda(t - t_k)}$$

$$= (\lambda h)^k e^{-\lambda t}.$$

To get the likelihood function $L(\lambda)$ one has to divide the last expression by h^k, but this is not necessary for our purpose which is to maximize L with respect to λ. It can be seen directly that the maximum is attained at $\hat{\lambda} = k/t$. Since k stands here for $N(t)$, we conclude that the MLE for λ is $\hat{\lambda} = N(t)/t$.

It is interesting to note that although we have the full information about the process N up to time t, only its value at time t matters for the MLE. This is explained by the fact that $N(t)$ is *sufficient* with respect to λ (will be defined later).

Method of moments (MOM). This method is suitable primarily for the case of $\mathbf{X} = (X_1, X_2, \ldots, X_n)$ of independent and identically distributed (i.i.d.) random variables. Let $\mu_k = E_\theta X_1^k$ be the kth moment of X_i ($k = 1, 2, \ldots$) and let $M_k = n^{-1}\Sigma_i X_i^k$ be the empirical kth moment. In order to estimate $\tau = g(\theta)$, one has first to express τ as a function of the moments, say $\tau = \psi(\mu_{i_1}, \mu_{i_2}, \ldots, \mu_{i_r})$. Then, by the MOM, the estimator of τ is $\hat{\tau} = \psi(M_{i_1}, M_{i_2}, \ldots, M_{i_r})$. In particular, the empirical moments M_k are the natural estimators of the corresponding theoretical moments μ_k. Note, the function ψ is not unique, so one can find more than one candidate for estimating the same τ. We will prefer to use the one with the smallest MSE. Another criterion could be the simplest, namely, the one with the smallest r. In many cases the two criteria lead to the same estimator.

We now apply this method to the last two examples.

Exponential sample. The kth moment of the exponential distribution is $\mu_k = k!/\lambda^k$. Suppose we want to estimate λ. Then, since $\lambda = 1/\mu_1$, the simplest MOM estimator is $\hat{\lambda} = 1/M_1 = 1/\bar{X}$, same as the MLE. Now, suppose we want to estimate the variance,

$$\sigma^2 = \text{Var} X_1 = E_\lambda\{X_1 - E_\lambda X_1\}^2 = \mu_2 - \mu_1^2 = \frac{1}{\lambda^2} = \mu^2.$$

Since the variance can be expressed either as μ_1^2 or $\mu_2 - \mu_1^2$, we have here two candidates, $\hat{\sigma}_1^2 = \bar{X}^2$ and $\hat{\sigma}_2^2 = M_2 - M_1^2 = n^{-1}\Sigma(X_i - \bar{X})^2$, the first is based on the sample mean, the second is the sample variance.

Poisson process. The method of moments is not designed for a continuous-time stochastic process. However, if we consider $N(t)$ as a sample of size 1, we can estimate λ. Here $\mu_1 = E_\lambda N(t) = \lambda t$, hence by this method the estimator for λ is $\hat{\lambda} = N(t)/t$, same as the MLE.

To see that the two methods do not always yield the same estimators, consider the following example.

Power distribution. Suppose we observe a sample of size n, namely $\mathbf{X} = (X_1, X_2, \ldots, X_n)$, where the X_i are independent with a common p.d.f. $f_\theta(x) = \theta x^{\theta-1}$ $(0 \leq x \leq 1)$. Here $\Theta = \{\theta : \theta > 0\}$. The likelihood function

$$L(\theta) = \prod_{i=1}^{n} \theta X_i^{\theta-1},$$

which attains its maximum at the MLE

$$\hat{\theta} = \frac{n}{-\sum \log X_i}.$$

Now, it is easy to evaluate $\mu_1 = \theta/(1+\theta)$, thus, $\theta = \mu_1/(1-\mu_1)$. Hence, the MOM estimator is

$$\theta^* = \frac{M_1}{1 - M_1} = \frac{\bar{X}}{1 - \bar{X}}.$$

1.3.3 Consistency

An estimator $\hat{\tau}$ of τ is said to be consistent if it converges to τ as more and more data are observed. More specifically, in the case of a sample $\mathbf{X} = (X_1, X_2, \ldots, X_n)$ of iid random variables, $\hat{\tau} = \hat{\tau}_n(X_1, X_2, \ldots, X_n)$ is *consistent if $\hat{\tau}_n \overset{p}{\to} \tau$ as $n \to \infty$*. In the case of a continuous time stochastic process $\mathbf{X} = \{X(s) : 0 \leq s \leq t\}$, $\hat{\tau} = \hat{\tau}_t(\mathbf{X})$ is consistent if $\hat{\tau}_t \overset{p}{\to} \tau$ as $t \to \infty$. Here $\overset{p}{\to}$ means convergence in probability, namely, for all $\epsilon > 0$,

$$\lim_{n \to \infty} P\{|\hat{\tau}_n - \tau| > \epsilon\} = 0.$$

By the Chebychev inequality, a sufficient condition for an estimator to be consistent is that its MSE tends to 0. Take for instance the exponential example. We have seen that the MLE for λ is $\hat{\lambda} = n/\Sigma X_i$. Knowing that $\Sigma X_i \sim \Gamma(n, \lambda)$, we find out that the bias is $\lambda/(n-1)$ and the variance is $(\lambda n)^2/((n-1)^2(n-2))$ $(n \geq 3)$ (see Problem 6 of this section). Hence,

$$MSE_\lambda(\hat{\lambda}) = \frac{\lambda^2(n+2)}{(n-1)(n-2)} \quad (n \geq 3),$$

which tends to 0 as $O(1/n)$. Another argument, often used, is as follows. Let $\tau = 1/\lambda$. Its MLE $\hat{\tau} = \bar{X}$ is unbiased with variance λ^{-2}/n. Hence, $\hat{\tau}$ is consistent, i.e., tends to τ (in probability). This implies that every continuous transformation $\phi(\hat{\tau})$ is consistent with respect to $\phi(\tau)$. The transformation $\phi : x \mapsto 1/x$ $(x > 0)$ is continuous, except at 0. But this causes no difficulty; if τ is arbitrarily close to 0, by the law of large numbers, for n large enough, $\bar{X}$ is also arbitrarily close to 0 and hence in this case $\hat{\lambda} = 1/\hat{\tau}$ is arbitrarily large.

1.3.4 Sufficiency

A statistic $S = S(\mathbf{X})$ is *sufficient* with respect to θ if the conditional distribution of $\mathbf{X}$ given S is θ-free. That is, if $f(\mathbf{x} \mid S(\mathbf{x}) = s)$ does not depend on θ. Take for instance $\mathbf{X} = (X_1, X_2, \ldots, X_n)$, n independent Poisson random variables with parameter λ. We shall see that $S(\mathbf{X}) = \Sigma X_i$ is sufficient. For that purpose compute

$$P\{X_1 = x_1, \ldots, X_n = x_n \mid S = s\} = \frac{P\{X_1 = x_1, \ldots, X_n = x_n, S = s\}}{P\{S = s\}}.$$

(1.32)

Now, the numerator of Equation (1.32) vanishes if $\Sigma x_i \neq s$. So, suppose $\Sigma x_i = s$. Then "$S = s$" is redundant in the numerator of Equation (1.32) and can be suppressed. The independence of the X_i implies that Equation (1.32) is equal to

$$\frac{\Pi e^{-\lambda} \lambda^{x_i} / x_i!}{e^{-n\lambda}(n\lambda)^s / s!} = \frac{s!}{n^s \Pi(x_i!)},$$

which is λ-free, thus proving that the sum of the observations, in this example, is sufficient.

A sufficient statistic carries all the information in the data about the unknown parameter θ. Once this statistic is calculated, there is nothing more we can learn about θ from the data. We note here that if S is sufficient, then any one-to-one transformation of S is also sufficient. Thus, in our last example, $\bar{X}$ is also sufficient.

An easy way to recognize a sufficient statistics is the *Neyman-Fisher factorization*. A statistic $S = S(\mathbf{X})$ is sufficient for $\theta \in \Theta$, if and only if, the p.d.f. of $\mathbf{X}$ can be factored as follows:

$$f_\theta(\mathbf{x}) = h(\mathbf{x}) \cdot \varphi_\theta(S(x)) \quad (\theta \in \Theta),$$

(1.33)

where $h(\mathbf{x})$ does not depend on θ and $\varphi_\theta(S(\mathbf{x}))$ depends on $\mathbf{x}$ only through $S(\mathbf{x})$.

For the Poisson sample, the factorization

$$f_\lambda(\mathbf{x}) = \frac{1}{\Pi(x_i!)} \cdot \lambda^{\Sigma x_i} e^{-n\lambda}$$

exhibits the idea. For the homogeneous Poisson process, observed from time 0 to time t, with failure-times t_i, the p.d.f. is nonzero if $0 < t_1 < t_2 < \ldots$. Incorporating this fact into the p.d.f. (which is derived in Equation (1.31)),

$$f_\lambda(\mathbf{x}) = \mathbf{1}\{0 < t_1 < t_2 < \cdots < t_k \leq t\} \cdot \lambda^k e^{-\lambda t}.$$

(1.34)

Here k stands for $N(t)$. This factorization proves that $N(t)$ is sufficient for λ. So, in this case, although we might have kept records of all the failure-times, for estimating λ, we only need to know the total number of failures up to time t.

A last word about the Neyman-Fisher factorization — since maximizing $f_\theta(\mathbf{x})$ is equivalent to maximizing $\varphi_\theta(S(\mathbf{x}))$, an MLE is always a function of a sufficient statistic!

1.3.5 Rao-Blackwell improved estimator

A clever use of a sufficient statistic is as follows. Suppose $\hat{\tau}(\mathbf{X})$ is an estimator of $\tau = g(\theta)$ and $S(\mathbf{X})$ is a sufficient statistic (with respect to $\theta \in \Theta$). The *Rao-Blackwell improved estimator* is

$$\tau^* = E(\hat{\tau} \mid S).$$

First, we note that τ^* is indeed a statistic, since the evaluation of the conditional expectation does not involve θ. Second,

$$E_\theta \hat{\tau} = E_\theta \tau^*,$$

hence both have the same bias (if one is unbiased, so is the other). Third,

$$\text{Var}_\theta \tau^* \leq \text{Var}_\theta \hat{\tau}. \qquad (1.35)$$

Equality of variances occurs only when $\hat{\tau}$ is already a function S. The reason for inequality (1.35) is that we can decompose $\text{Var}\,\hat{\tau}$ into two terms

$$\text{Var}\,\hat{\tau} = E\{\text{Var}(\hat{\tau}|S)\} + \text{Var}\{E(\hat{\tau}|S)\}$$

and the second term on the right is $\text{Var}\,\tau^*$.

To demonstrate its usefulness consider again the homogeneous Poisson process N, observed from time 0 to time t. We already know that $N(t)$ is sufficient with respect to the rate parameter λ. Let t_0 ($0 < t_0 < t$) be fixed and suppose we want to estimate $\tau = P\{N(t_0) = 0\} = e^{-\lambda t_0}$. The maximum likelihood and moments methods, both will first estimate λ by $\hat{\lambda} = N(t)/t$ and then substitute it in $e^{-\lambda t_0}$ to get $e^{-\hat{\lambda} t_0}$. Note, this is a biased estimator of τ. The Rao-Blackwell approach is to choose a trivial unbiased estimator and then to improve it. Let

$$\hat{\tau} = \mathbf{1}\{N(t_0) = 0\}.$$

This trivial estimator, which takes on only the values 0 and 1, is unbiased. Its improved version is

$$\tau^* = E(\hat{\tau} \mid N(t)) = P\{N(t_0) = 0 \mid N(t)\} = \left(1 - \frac{t_0}{t}\right)^{N(t)}. \qquad (1.36)$$

We shall see that this estimator has the smallest possible variance among unbiased estimators.

1.3.6 Complete statistic

A statistic $S(\mathbf{X})$ is *complete* with respect to $\theta \in \Theta$ if

$$E_\theta \, h(S(\mathbf{X})) = 0 \quad (\theta \in \Theta),$$

necessarily implies that $h \equiv 0$. This is in fact a property of the family of distributions $\{f_\theta : \theta \in \Theta\}$. Now, suppose $S(\mathbf{X})$ is complete and two unbiased estimators $\hat{\tau}_1(S)$, $\hat{\tau}_2(S)$ are proposed to estimate $\tau = g(\theta)$. It follows that

$$E_\theta \{\hat{\tau}_1(S) - \hat{\tau}_2(S)\} = 0 \quad (\theta \in \Theta).$$

The completeness implies that $\hat{\tau}_1 \equiv \hat{\tau}_2$, i.e., there is in fact only one unbiased estimator of τ which is based on the complete statistic S. If $S(\mathbf{X})$ is also sufficient, then $\hat{\tau}_1(S)$ cannot be improved and is therefore the (unique) *uniformly minimum variance unbiased estimator* (UMVUE) of τ. Here, the word *uniformly* emphasizes the fact that the *minimum variance* property holds whatever θ may be.

Looking again at the homogeneous Poisson process, let us check whether $N(t)$ is complete. Since t is fixed, $Y = N(t)$ is just a Poisson random variable with mean $\mu = \lambda t$. Let h be a real function defined on the natural numbers and suppose that

$$E_\mu h(Y) = e^{-\mu} \sum_{j=0}^{\infty} \frac{h(j)}{j!} \mu^j = 0 \quad (\mu > 0).$$

The factor $e^{-\mu}$ in front of the sum is always positive, thus we have a power series which vanishes on $(0, \infty)$. This is possible if and only if all the coefficients vanish, namely, $h \equiv 0$. Hence, the Poisson distribution is complete. It follows, in particular, that the estimator of $e^{-\lambda t_0}$ which is given in Equation (1.36) is UMVUE.

1.3.7 Confidence intervals

All the estimators we have dealt with so far are *point-estimators*. An *interval-estimator* is determined by two statistics $L(\mathbf{X})$ and $U(\mathbf{X})$ $(L < U)$. Suppose we want to estimate $\tau = g(\theta)$ by an interval $[L, U]$ such that the probability that the interval covers τ is no less than a pre-assigned level $1 - \alpha$ (e.g., $1 - \alpha = .95$), namely,

$$P_\theta\{L \le \tau \le U\} \ge 1 - \alpha \quad (\theta \in \Theta). \tag{1.37}$$

In many cases, we can find statistics U and L that satisfy condition (1.37) with equality. This coverage probability is called *confidence-level*. Naturally, the shorter the interval $[L, U]$ the estimation of the parameter is more accurate but less confident.

Given a confidence-level $1-\alpha$, we first choose a point-estimator $\hat{\tau}$ for τ, preferably one which is based on a sufficient statistic. Then, we make a convenient change of variable $Q = Q(\hat{\tau})$, such that Q has a distribution which is θ-free. As it turns out, the result is a random variable Q which depends on $\hat{\tau}$ and on θ and is called the *pivotal function*. The third step is to choose two numbers p_1, p_2 in $[0, 1]$ such that $p_1 - p_2 = 1 - \alpha$ and calculate (or read from a table) the corresponding quantiles of Q, $q(p_1)$ and $q(p_2)$. It follows that

$$P\{q(p_2) \leq Q \leq q(p_1)\} = 1 - \alpha.$$

The last step is to translate the two inequalities for Q into inequalities for τ.

When Q has a symmetric distribution (or nearly symmetric), we take $p_1 = 1 - \alpha/2$ and $p_2 = \alpha/2$. Otherwise (if we can numerically) we choose p_1 so as to make the interval as short as possible. We demonstrate the procedure through two famous examples.

Normal sample. Suppose $\mathbf{X} = (X_1, X_2, \ldots, X_n)$, where the X_i are independent, identically distributed random variables (i.i.d.), $X_i \sim N(\mu, \sigma^2)$. Suppose σ is known and we want to construct a .95-confidence interval for μ. The reader can verify that here $S(\mathbf{X}) = \Sigma X_i$ is sufficient and $\hat{\mu} = \bar{X} = S/n$ is both MLE and moments estimator. We transform $\bar{X}$ into $Q = n^{1/2}(\bar{X} - \mu)/\sigma$, the latter is a standard normal random variable (which is symmetric). From the Standard-Normal-Table we extract $q(.975) = 1.96 = -q(.025)$. We then write

$$-1.96 \leq n^{1/2}(\bar{X} - \mu)/\sigma \leq 1.96$$

which is equivalent to

$$\bar{X} - 1.96\sigma/n^{1/2} \leq \mu \leq \bar{X} + 1.96\sigma/n^{1/2}.$$

Here $L = \bar{X} - 1.96\sigma/n^{1/2}$ and $U = \bar{X} + 1.96\sigma/n^{1/2}$.

Exponential sample. Suppose now that the sample is from an exponential distribution with parameter λ and we want to construct a .9-confidence interval for λ. We already know that $S = \Sigma X_i$ is sufficient and the MLE is $\hat{\lambda} = n/S$. Let $Q = n\lambda/\hat{\lambda} = \lambda S$, then $Q \sim \Gamma(n, 1)$. Now, for moderately large n, the gamma distribution is nearly symmetric and we extract the quantiles $q(.95)$ and $q(.05)$ from a Gamma-Table (or calculate with the help of some computer software as *S-PLUS* or *Mathematica*). The next step is to write

$$q(.05) \leq \lambda S \leq q(.95)$$

which is equivalent to

$$q(.05)/S \leq \lambda \leq q(.95)/S.$$

Since a Chi-Square-Table is more accessible, one might prefer to use the latter. In this case the pivotal function is $Q_2 = 2\lambda S$ which is a chi-square random variable with $2n$ degrees of freedom.

Uniform sample. Suppose the sample $\mathbf{X} = (X_1, X_2, \ldots, X_n)$ is from the uniform distribution in $[0, \theta]$, $\theta > 0$. Then, the largest observation X_{max} is sufficient and MLE for θ. Define $Q = X_{max}/\theta$, then $q(p) = p^{1/n}$, $0 \leq p \leq 1$. Given a confidence-level $1 - \alpha$, the shortest interval is obtained for the choice $p_1 = 1$, $p_2 = \alpha$. The inequalities

$$\alpha^{1/n} \leq \frac{X_{max}}{\theta} \leq 1$$

lead to the confidence interval

$$X_{max} \leq \theta \leq \frac{X_{max}}{\alpha^{1/n}}.$$

1.3.8 Order statistics

Let $X_1, X_2, \ldots, X_n$ be a sample of size n from a (cumulative) distribution function F. Let

$$X_{(1)} \leq X_{(2)} \leq \cdots \leq X_{(n)}$$

be the ordered X_i. The random variable $X_{(k)}$ is called the kth *order statistic*. Let $J(x) = \Sigma \mathbf{1}\{X_i \leq x\}$ be the number of sample-values which are less or equal to x and we note that $J(x) \sim B(n, F(x))$. Since the events $\{X_{(r)} \leq x\}$ and $\{J(x) \geq r\}$ are equivalent, the distribution function of $X_{(r)}$ is given by

$$F_r(x) := P\{X_{(r)} \leq x\} = \sum_{i=r}^{n} \binom{n}{i} F^i(x)(1 - F(x))^{n-i} \quad (1 \leq r \leq n).$$

$$(1.38)$$

If F is absolutely continuous with p.d.f. $f = F'$, then $f_r = F'_r$ is given by

$$f_r(x) = \frac{n!}{(r-1)!(n-r)!} F^{r-1}(x)(1 - F(x))^{n-r} f(x) \quad (1 \leq r \leq n).$$

$$(1.39)$$

In particular, the c.d.f.s of the smallest and largest order statistics are given by

$$F_1(x) = 1 - (1 - F(x))^n, \qquad F_n(x) = F^n(x).$$

Note, $X_{(r)} \sim \text{beta}(r, n+1-r)$ when $F(x) = x$ $(0 \leq x \leq 1)$ is the uniform distribution.

Joint distributions are also easy to get in a closed form. For instance let us compute the joint p.d.f. $f_{r,s}$ of $(X_{(r)}, X_{(s)})$ for $1 \leq r < s \leq n$. Defining $J(x, y) = J(y) - J(x)$ for $x < y$, we argue as follows. Let $x < y$ and let $h > 0$ be small $(h < y - x)$. Then

$$P\{x - h < X_{(r)} \leq x, y - h < X_{(s)} \leq y\} \qquad (1.40)$$
$$= P\{J(x - h) = r - 1, J(x - h, x) = 1, J(x, y - h) = s - r - 1,$$
$$J(y - h, y) = 1, J(y, \infty) = n - s\} + o(h^2).$$

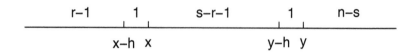

Figure 1.2 *Numbers of sample-values in each interval.*

The $o(h^2)$ stands for the probability that at least two of the X_i fall in $(x-h, x]$ or in $(y-h, y]$ (see Figure 1.2).

Noting that $P\{x - h < X_i \leq x\} = f(x)h + o(h)$, and that the 5-dimensional random vector $(J(x-h), J(x-h,x), J(x,y-h), J(y-h,y), J(y,\infty))$ has a multinomial distribution, Equation (1.40) is equal to

$$\frac{n!}{(r-1)!1!(s-r-1)!1!(n-s)!}F^{r-1}(x-h) \cdot f(x)h \tag{1.41}$$
$$\cdot \{F(y-h) - F(x)\}^{s-r-1} \cdot f(y)h \cdot \{1 - F(y)\}^{n-s} + o(h^2).$$

Dividing by h^2 and taking the limit as $h \downarrow 0$, we finally get

$$f_{r,s}(x,y) = \frac{n!}{(r-1)!(s-r-1)!(n-s)!}F^{r-1}(x)\{F(y) - F(x)\}^{s-r-1}$$
$$\cdot \{1 - F(y)\}^{n-s}f(x)f(y) \qquad (1 \leq r < s \leq n; \; x \leq y). \tag{1.42}$$

Similar arguments can be used to derive the joint density function of any number of the order statistics. We shall just say in closing that the joint density of all the order statistics is given by

$$f_{1,2,\ldots,n}(x_1, x_2, \ldots, x_n) = n! \prod_{i=1}^{n} f(x_i) \qquad (x_1 \leq x_2 \leq \cdots \leq x_n). \tag{1.43}$$

An interesting property of the homogeneous Poisson process $N(t)$ is related to order statistics from a uniform distribution. Using the notation and conditions of Equation (1.31), it follows that the conditional joint density of the failure-times $0 < T_1 < T_2 < \cdots < T_k \leq t$, given the event "$N(t) = k$" is given by

$$f_{1,2,\ldots,k}(t_1, t_2, \ldots, t_k) = \frac{\lambda^k e^{-\lambda t}}{(\lambda t)^k e^{-\lambda t}/k!} = \frac{k!}{t^k}. \tag{1.44}$$

This is a special case of Equation (1.43), where $f(x) = 1/t$ $(0 \leq x \leq t)$ is the p.d.f. of the uniform distribution on $[0, t]$.

1.4 Generating a Poisson process

For simulation purposes, one might be interested in generating a homogeneous Poisson process. We have seen in Section 1.2 that the time intervals between successive occurrences in a homogeneous Poisson process with constant event rate λ per unit time are mutually independent random variables, each possessing the same exponential p.d.f. $f(t) = \lambda e^{-\lambda t}$, $t \geq 0$. This is a key property of the homogenous Poisson process and provides a way of generating sample functions of the Poisson process. More precisely, let $X_1, X_2, \ldots, X_k, \ldots$ be independent observations all drawn from a common exponential density function $f(x) = \lambda e^{-\lambda x}$, $x \geq 0$. Let us further define the ordered sequence of times: $T_1 = X_1, T_2 = T_1 + X_2, T_3 = T_2 + X_3, \ldots, T_k = T_{k-1} + X_k, \ldots$. It can then be shown that the step function defined as 0 in the interval $0 \leq t < T_1$, 1 in the interval $T_1 \leq t < T_2, \ldots, j$ in the interval $T_j \leq t < T_{j+1}, \ldots$, is distributed the same as a homogeneous Poisson process having occurrence rate λ. Indeed, we shall prove that interarrival times $\{X_i\}$ having a common exponential density function give rise to a discrete $N(t)$ which follows the Poisson distribution (1.12) for each fixed t.

To prove this, we first note that, for any preassigned positive time $t > 0$, the event $N(t) \geq k$ is completely equivalent to the event $T_k = X_1 + X_2 + \cdots + X_k \leq t$, i.e.,

$$P(N(t) \geq k) = P(T_k \leq t). \qquad (1.45)$$

Hence

$$
\begin{aligned}
P(N(t) = k) &= P(N(t) \geq k) - P(N(t) \geq k+1) \qquad (1.46)\\
&= P(T_k \leq t) - P(T_{k+1} \leq t) = F_k(t) - F_{k+1}(t),
\end{aligned}
$$

where $F_k(t)$ and $F_{k+1}(t)$ are, respectively, the c.d.f.s of T_k and T_{k+1}. These formulae hold for all integers $k \geq 1$. For $k = 0$, we add that $P(N(t) = 0) = P(T_1 > t) = \int_t^\infty \lambda e^{-\lambda \tau} d\tau = e^{-\lambda t}$.

If U and V are independent nonnegative random variables with p.d.f.s $f(u)$ and $g(v)$, respectively, then the random variable $W = U + V$ is distributed with p.d.f. $h(w)$, which is derived via the *convolution formula* $h(w) = \int_0^w f(u) g(w-u) du$. We denote this by $h = f * g$. Here, $T_k = T_{k-1} + X_k$, where T_{k-1} and X_k are independent nonnegative random variables with p.d.f.s $f_{k-1}(t)$ and $\lambda e^{-\lambda x}$, respectively. Hence $f_k(t)$, the p.d.f. of T_k, satisfies the relation $f_k(t) = \int_0^t \lambda e^{-\lambda x} f_{k-1}(t-x) dx$. From this, it follows by induction that $f_j(t)$, $j = 1, 2, \ldots$, is the Erlang p.d.f. $\lambda(\lambda t)^{j-1} e^{-\lambda t}/(j-1)!$, $t \geq 0$. The associated c.d.f. is

$$F_j(t) = \int_0^t f_j(\tau) d\tau = \sum_{i=j}^{\infty} \frac{e^{-\lambda t}(\lambda t)^i}{i!}.$$

Substituting in (1.46) we get

$$P(N(T) = k) = \sum_{i=k}^{\infty} \frac{e^{-\lambda t}(\lambda t)^i}{i!} - \sum_{i=k+1}^{\infty} \frac{e^{-\lambda t}(\lambda t)^i}{i!}$$

$$= \frac{e^{-\lambda t}(\lambda t)^k}{k!}. \qquad (1.47)$$

Thus, for every fixed $t > 0$, $N(t)$ follows the Poisson distribution. To prove that $\{N(t) : t \geq 0\}$ is a homogeneous Poisson process, one has to prove that all finite-dimensional distributions are also of a homogeneous Poisson process. That is, for every $k \geq 2$, integers $j_0 = 0 \leq j_1 \leq j_2 \leq \cdots \leq j_k$ and time points $t_0 = 0 < t_1 < t_2 < \cdots < t_k$, the probability

$$P\left\{ \bigcap_{i=1}^{k} \{N(t_i) = j_i\} \right\} = P\left\{ \bigcap_{i=1}^{k} \{T_{j_i} \leq t_i < T_{j_i+1}\} \right\}$$

is the same as the corresponding one for a homogeneous Poisson process. Indeed, the right-hand side is the desired one.

It is interesting to raise the question: What is the distribution of $N(t)$, the number of occurrences up to time t, if the interarrival times $\{X_i\}$ are mutually independent but follow a general c.d.f. $F(x)$? It is easily seen that Equations (1.45) and (1.46) hold in general, i.e.,

$$P(N(t) \geq k) = F_k(t) \qquad (1.48)$$

and

$$P(N(t) = k) = F_k(t) - F_{k+1}(t), \qquad k = 0, 1, 2, \ldots, n \qquad (1.49)$$

where $F_k(t)$ and $F_{k+1}(t)$ are, respectively, the c.d.f.s of $T_k = X_1 + X_2 + \cdots + X_k$ and $T_{k+1} = X_1 + X_2 + \cdots + X_{k+1}$ and where

$$F_0(t) = 1, \qquad t \geq 0$$
$$= 0, \qquad t < 0.$$

Since the $\{X_i\}$ are mutually independent nonnegative random variables with common c.d.f. $F(x)$, the recursion formula $F_{k+1}(t) = \int_0^t F_k(t - \tau)dF(\tau)$, (i.e., $F_{k+1} = F_k * F$) $k = 0, 1, 2, \ldots$, can be used to find the c.d.f.s of $T_{k+1} = \sum_{i=1}^{k+1} X_i$. If $F'(t) = f(t)$ exists, then the p.d.f. $f_{k+1}(t) = \int_0^t f_k(t - \tau)f(\tau)d\tau, k = 1, 2, \ldots,$. For more details, see Chapter 9, Example 1 of Section 9.2.

1.5 Nonhomogeneous Poisson process

A useful generalization of the homogeneous Poisson process with constant event rate λ is the time-dependent Poisson process with event rate $\lambda(t)$, also known as the *intensity function*. The assumptions for the time-dependent Poisson process are precisely the same as for the homogeneous case with λ replaced by $\lambda(t)$ in assumption (ii).

For the time-dependent Poisson process, the derivations for $P_n(t) = P[N(t) = n]$ can be carried out in a manner precisely analogous to the one used for the homogeneous case. It can be verified that

$$P_n(t) = e^{-\Lambda(t)}[\Lambda(t)]^n/n!, \qquad n = 0, 1, 2, \ldots, \qquad (1.50)$$

where $\Lambda(t) = \int_0^t \lambda(\tau)d\tau$, i.e., $N(t)$ is a Poisson random variable with parameter $\Lambda(t)$. Furthermore, $E[N(t)] = \text{Var}[N(t)] = \Lambda(t)$. The function Λ can be thought of as a measure on $[0, \infty)$, namely, the measure of an interval $[a, b]$ is $\Lambda([a, b]) := \Lambda(b) - \Lambda(a) = E\, N(a, b)$. Thus, Λ is called the *mean measure* of the Poisson process.

It can also be verified that the p.d.f. and c.d.f. of T_k, the time to the kth occurrence are given by

$$f_k(t) = \lambda(t)[\Lambda(t)]^{k-1}e^{-\Lambda(t)}/(k-1)!, \qquad t \geq 0 \qquad (1.51)$$

and

$$F_k(t) = 1 - \sum_{j=0}^{k-1} e^{-\Lambda(t)}\frac{[\Lambda(t)]^j}{j!}, \qquad t \geq 0. \qquad (1.52)$$

In particular, the p.d.f. and c.d.f. of T_1, the time to the first occurrence, are given by

$$f_1(t) = \lambda(t)e^{-\Lambda(t)}, \qquad t \geq 0 \qquad (1.53)$$

and

$$F_1(t) = 1 - e^{-\Lambda(t)}, \qquad t \geq 0. \qquad (1.54)$$

Equations (1.51) and (1.53) become, respectively, the exponential and gamma distributions for the case where $\lambda(t) \equiv \lambda$. It should be emphasized, however, that many of the results which are true for a homogeneous Poisson process are no longer valid for a time-dependent Poisson process.

Thus, for example, for a homogeneous Poisson process with event rate λ the distribution of $N(t, t + \tau) = N(t + \tau) - N(t)$, the number of occurrences in the time interval $(t, t + \tau]$, is independent of the starting point t and depends only on τ. More precisely, the distribution of $N(t, t + \tau)$ is given by

$$P[N(t, t + \tau) = k] = \frac{e^{-\lambda\tau}(\lambda\tau)^k}{k!}, \qquad k = 0, 1, 2, \ldots. \qquad (1.55)$$

For a time-dependent Poisson process with time-dependent event rate $\lambda(t)$, the distribution of $N(t + \tau) - N(t)$ does depend on the location of the interval

$(t, t + \tau]$ and is, in fact, given by

$$P[N(t, t + \tau) = k] = \frac{e^{-[\Lambda(t+\tau)-\Lambda(t)]}[\Lambda(t + \tau) - \Lambda(t)]^k}{k!}, \quad k = 0, 1, 2, \ldots.$$

(1.56)

As another example, we recall that for a homogeneous Poisson process times between successive occurrences are mutually independent and distributed with common exponential p.d.f. $\lambda e^{-\lambda t}$, $t \geq 0$. This is not true for the time-dependent Poisson process. More precisely, if T_j represents the time of the jth occurrence (time measured from $t = 0$), then the interarrival times $T_1, T_2 - T_1, T_3 - T_2, \ldots, T_j - T_{j-1}, \ldots$ are neither mutually independent nor identically distributed. Thus, if the jth interarrival time $T_j - T_{j-1}$ is denoted by the random variable Y_j, it can be shown that

$$P(Y_j > \tau) = \int_0^\infty e^{-\Lambda(t+\tau)} \lambda(t) \frac{[\Lambda(t)]^{j-2}}{(j-2)!} \, dt.$$

(1.57)

Only for the special case of a homogeneous Poisson process where $\lambda(t) \equiv \lambda$ for all t does it follow that $P(Y_j > \tau) = e^{-\lambda\tau}$ independently of j.

Suppose we want to generate a time-dependent Poisson process with a given intensity function $\lambda(t)$ and its associated mean measure Λ. We start with generating a sequence of i.i.d. unit-exponential random variables $\{X_i\}$ and define the sequence $\{S_j = \Sigma_{i=1}^j X_i : j = 1, 2, \ldots\}$ which is the event-times of a standard (i.e., $\lambda = 1$) homogeneous Poisson process. Let $T_j = \Lambda^{-1}(S_j)$, then $\{T_j : j = 1, 2, \ldots\}$ are the event-times of the desired Poisson process and if $N(t)$ is the number of T_j in $[0, t]$ then $\{N(t) : t \geq 0\}$ is the process itself.

The occurrence over time of failures in computers, automobiles, aircraft, satellites, human beings, and other complex systems can be considered as a time-dependent Poisson process. In reliability applications $\lambda(t)$ is the (instantaneous) rate at which failures occur at time t and $\Lambda(t) = \int_0^t \lambda(\tau) d\tau$ is the integrated failure rate over the interval $(0, t]$. Thus, $\Lambda(t)/t$ is the *average failure rate over the time interval* $(0, t]$. It should be pointed out, however, that although $\Lambda(t)/t = \lambda t/t = \lambda$ in the particular case of a homogeneous Poisson process with constant failure rate λ, it is not true in general that $\Lambda(t)/t$ equals $\lambda(t)$. For example, if $\lambda(t) = \lambda(\lambda t)^{\beta-1}$, then $\Lambda(t)/t = \lambda(\lambda t)^{\beta-1}/\beta = \lambda(t)/\beta$. Thus $\Lambda(t)/t = \lambda(t)$ if $\beta = 1$ (constant failure rate); $\Lambda(t)/t > \lambda(t)$, if $0 < \beta < 1$ [$\lambda(t)$ is monotonically decreasing for values of β in this range]; and $\Lambda(t)/t < \lambda(t)$, if $\beta > 1$ [$\lambda(t)$ is monotonically increasing for values of β in this range].

In reliability literature $N(t)/t$, where $N(t)$ is the observed number of failures in $(0, t]$, is often given as an estimate of the instantaneous failure rate at time t. But since $EN(t)/t = \Lambda(t)/t$, $N(t)/t$ gives an unbiased estimate of the *average failure rate over* $(0, t]$. If the Poisson process is homogeneous, then $N(t)/t$, the average number of failures in unit-time observed in $(0,t]$, is the unbiased estimate of the constant instantaneous failure rate λ. It should be clear, however, that for a time-dependent Poisson process $N(t)/t$ will in general not give an unbiased estimate of the instantaneous failure rate $\lambda(t)$ but may, in fact, overestimate or underestimate $\lambda(t)$.

1.6 Three important discrete distributions

The binomial, geometric and negative binomial distributions are three important discrete distributions. The stochastic framework underlying each of these distributions is a sequence of independent trials, where each trial has only two possible outcomes, which we denote for convenience as success S and failure F, and where the probability of success, p, and probability of failure, $q = 1-p$, remain the same for all trials. We define the random variables X_i as the indicators of the success in the ith trial, namely, $X_i = 1$ for success and $X_i = 0$ for failure. We shall refer to such trials as the Bernoulli trials in honor of the early pioneer in probability theory, James Bernoulli.

Binomial distribution. Consider a sequence of n Bernoulli trials. Let us denote the number of successes by the random variable $S_n = X_1 + X_2 + \cdots + X_n$. The random variable S_n may assume any one of the integer values $k = 0, 1, 2, \ldots, n$. The associated probability distribution is

$$P(S_n = k) = \binom{n}{k} p^k q^{n-k}, \qquad k = 0, 1, 2, \ldots, n. \qquad (1.58)$$

Equation (1.58) is known as the *binomial distribution*. It can be verified that the mean and variance of S_n are, respectively,

$$E(S_n) = np \quad \text{and} \quad \text{Var}(S_n) = npq.$$

The case $n = 1$ ($S_1 = X_1$) is called Bernoulli distribution.

Geometric distribution. Again consider a sequence of Bernoulli trials and let us define the random variable Y_1 as that trial on which the first success occurs. It is clear that Y_1 is a discrete random variable, which may assume any one of the integer values $1, 2, 3, \ldots$. Since the event $Y_1 = j$ can occur if and only if the first $(j-1)$ Bernoulli trials are failures and the jth Bernoulli trial is a success, it follows at once that

$$P(Y_1 = j) = pq^{j-1}, \qquad j = 1, 2, \ldots. \qquad (1.59)$$

Equation (1.59) is known as the *geometric distribution*. It can be verified that the mean and variance of Y_1 are, respectively,

$$E(Y_1) = 1/p \quad \text{and} \quad Var(Y_1) = q/p^2.$$

Negative binomial distribution. This is a generalization of the geometric distribution. The random variable of interest is Y_k, the trial on which the kth success occurs, is a discrete random variable which may assume any one of the integer values $k, k+1, k+2, \ldots$. Since the event $Y_k = j$ can occur if and only if the two independent events A, "$(k-1)$ successes in the first $j-1$ trials" and B, "success on the jth trial" both occur, it follows that

$$P(Y_k = j) = P(AB) = P(A)\,P(B) = \left[\binom{j-1}{k-1}p^{k-1}q^{j-k}\right] \cdot p$$

$$= \binom{j-1}{k-1}p^k q^{j-k}, \qquad j = k, k+1, \ldots. \qquad (1.60)$$

Equation (1.60) is known as the *negative binomial* or *Pascal distribution*. The mean and variance of Y_k are, respectively,

$$E(Y_k) = k/p \quad \text{and} \quad \mathrm{Var}(Y_k) = kq/p^2\,.$$

The reader will notice that $EY_k = kE\,Y_1$ and $\mathrm{Var}\,Y_k = k\mathrm{Var}\,Y_1$, the reason being that the time of the kth success is a sum of k independent geometric random variables.

It is worth noting that there is an analogy between the homogeneous binomial process and its associated distributions (binomial, geometric, Pascal) on the one hand and the homogeneous Poisson process and its associated distributions (Poisson, exponential, gamma) on the other. By the homogeneous binomial process, we mean simply a discrete time stochastic process which arises when Bernoulli trials are performed at the discrete times $t = 1, 2, 3, \ldots$. The modifier "homogeneous" refers to the assumption that the probability of success remains the same for each trial.

The binomial distribution is the probability law of the number of successes, k, when we observe the binomial process up to time $t = n$, where n is a preassigned number of Bernoulli trials. Analogously, the Poisson distribution is the probability law of the number of occurrences, k, of some event (such as emissions, failures, etc.) when we observe the Poisson process for a preassigned length of time t.

The discrete waiting time distributions (geometric and Pascal) arise when we ask how many trials, n, are needed (or if trials occur one second apart, how many seconds are needed) for the first success or more generally for the kth success to occur in a homogeneous binomial process. Analogously, the continuous waiting time distributions (exponential and gamma) arise when we ask for the distribution of the length of time t until the first or more generally the kth occurrence of some event of interest.

It should also be clear that the geometric and Pascal distributions bear the same relation to the binomial distribution as do the exponential and gamma distributions to the Poisson distribution. For the Poisson process, the relationship between the c.d.f. of the continuous random variable T_k and the c.d.f. of the discrete random variable $N(t)$ is a consequence of the equivalence of the events $T_k \le t$ and $N(t) \ge k$.

For the binomial process, the relationship between the c.d.f. of the discrete random variable Y_k and the c.d.f. of the discrete random variable S_j is a consequence of the equivalence of the events $Y_k \le k$ and $S_j \ge k$ (clearly this is so since the kth success occurs on or before the jth trial if and only if there are at least k successes in the first j trials). Hence, it follows that

$$P(Y_k \le j) = P(S_j \ge k) = \sum_{s=k}^{j} \binom{j}{s} p^s q^{j-s} = 1 - \sum_{s=0}^{k-1} \binom{j}{s} p^s q^{j-s}. \quad (1.61)$$

In the special case for which $k = 1$, Equation (1.61) becomes

$$P(Y_1 \le j) = P(S_j \ge 1) = 1 - P(S_j = 0) = 1 - q^j. \quad (1.62)$$

1.7 Problems and comments

In Sections 1.1–1.3 we are assuming a homogeneous Poisson process denoted by $\{N(t)\}$ and the kth event-time is denoted by T_k.

Problems for Section 1.1

1. Verify the fact that $\sum_{n=0}^{\infty} P_n(t) = 1$.

2. Fill in the steps left out in Equations (1.14) and (1.15).

3. Prove that for large λt, the random variable $N(t)$ is approximately normally distributed with mean $= \lambda t$ and variance $= \lambda t$.

4. Suppose that the times of occurrence of successive failures in a computer can be thought of as arising from a homogeneous Poisson process with failure rate $\lambda = .01$/hour.

   (a) What is the probability that the computer will be failure free for 50 hours?

   (b) What is the probability that at most three failures will occur in a 200-hour interval?

   (c) What is the probability that more than 120 failures occur in a 10,000-hour time interval?
   Hint: Use Problem 3 in finding the answer to 4(c).

5. During a 20-year period 240 people committed suicide by jumping off a certain bridge. Estimate the probability that a month will go by without a suicide. What is the probability that the number of suicides in a month exceeds three?

6. The α particles given off by some radioactive material are counted over a 10,000-second time interval. In this time interval 2,500 α particles were detected by the Geiger counter. What is the probability that at least two α particles will be detected in a 10-second time interval?

7. The average number of calls coming into a certain switchboard during 3:00 p.m. and 4:00 p.m. (based on evidence collected over a long period of time) is 180. Assuming that calls are equally likely to appear at any time between 3 and 4 o'clock, what is the probability that there will be no incoming calls in the 60 seconds between 3:20 and 3:21? What is the probability that at least two calls are received in each of the three 60-second intervals—3:10 to 3:11, 3:20 to 3:21, 3:30 to 3:31?

8. The failure-rate of a certain type of transistor is assumed to be a constant $\lambda = 10^{-6}$ / hour. An electronic system contains 10,000 of these transistors. Failed transistors are replaced immediately by other good transistors.

   (a) What is the expected number of transistor failures in 10,000 hours of usage?

   (b) How many spares should be ordered as replacements for failed transistors so as to keep the probability of running out of spares $\leq .001$?

   (c) Answer (b) if we want enough spares for four systems, each being used for 10,000 hours.

9. A homogeneous Poisson process with constant occurrence rate λ is observed for a fixed length of time t. Suppose that there are $n \geq 1$ occurrences in $(0, t]$, i.e., $N(t) = n$. Verify that the conditional probability $P[N(t^*) = k \mid N(t) = n]$, where $0 < t^* < t$, is given by the binomial distribution $B(n, t^*/t)$. Verify that $E[N(t^*) \mid N(t) = n] = nt^*/t$ and $\text{Var}[N(t^*) \mid N(t) = n] = nt^*(1 - t^*)/t^2$.
   Hint:

$$P[N(t^*) = k \mid N(t) = n]$$

$$= \frac{P[N(t^*) = k] \cdot P[N(t^*, t) = n - k]}{P[N(t) = n]}.$$

10. The failures generated by a computer over time (instantaneous repair is assumed) are observed for $t = 1,000$ hours. In the first 200 hours, 7 failures are observed; and in the next 800 hours, 3 failures are observed. Is it reasonable, on the basis of this evidence, that computer failures are generated over time by a homogeneous Poisson process over the time interval $0 \leq t \leq 1,000$?

11. A record is kept of the occurrence of failures of a certain machine over time. In 10,000 hours of operation, 100 failures are observed with 32 failures occurring during the first 2,000 hours. On the basis of the above evidence, would you be willing to assert that the failures can be thought of as being

generated by a Poisson process with constant failure rate over the entire 10,000-hour interval? Sketch the statistical reasoning.

Problems for Section 1.2

1. (a) Verify that $\int_0^\infty f_k(t)dt = 1$.

   (b) Verify the formulae in Equation (1.29).

      Hint: Verify that $\Gamma(n+1) := \int_0^\infty x^n e^{-x} dx = n!$, for any integer n.

2. Verify Equation (1.18) directly by computing $\int_0^t f_k(x)dx$.

3. Use Equation (1.34) to prove the assertion that the random variables $\{Y_j\}$, where $Y_1 = T_1, Y_2 = T_2 - T_1, \ldots, Y_j = T_j - T_{j-1}, \ldots$ are mutually independent with common p.d.f. $\lambda e^{-\lambda t}$, $t \geq 0$.

4. Prove that the random variable $2\lambda T_k$ is distributed as chi-square with $2k$ degrees of freedom [denoted as $\chi^2(2k)$].

   Hint: The p.d.f. of a random variable X distributed as chi-square with ν degrees of freedom is

$$f(x) = \frac{(2x)^{\nu/2-1}e^{-x/2}}{2\Gamma(\nu/2)}, \qquad x \geq 0.$$

5. Prove that for large k, the random variable T_k is approximately normally distributed with mean $= k\theta$ and variance $= k\theta^2$.

6. Suppose that customers arrive at a checkout station according to a Poisson process with rate $\lambda = 1/2$ (mean time between arrivals = 2 minutes). What is the probability that it takes more than 10 minutes for the fourth customer to arrive?

7. Suppose that the occurrence of successive failures for some systems can be adequately described by a homogeneous Poisson process with MTBF, $\theta = 500$ hours.

  (a) What is the probability that exactly 3 failures occur in the first 1,000 hours? At most 3 failures? At least 3 failures?

  (b) What is the probability that the third failure occurs before 1,000 hours?

  (c) Find that time t^* such that $P(T_3 > t^*) = .05$.

  (d) Compute the probability that the time to the 16th failure exceeds 10,000 hours.

  (e) Find that time t^* such that $P(T_{25} \leq t^*) = .99$.

8. A beta distribution with parameters $r, s > 0$ is defined by its p.d.f. $f(x) = B(r,s)x^{r-1}(1-x)^{s-1}$, $0 \leq x \leq 1$, where $B(r,s) = \Gamma(r+s)/(\Gamma(r)\Gamma(s))$.

A homogeneous Poisson process with constant occurrence rates λ is observed for a fixed length of time t. Given that $N(t) = n$, prove that the conditional distribution of T_k/t is beta$(k, n - k + 1)$, $1 \leq k \leq n$. Use this to compute the conditional mean and variance of T_k.

9. Prove that for $1 \leq k < n$, the conditional distribution of T_k/T_n, given T_n, is beta$(k, n - k)$. Use this to compute the conditional mean and variance of T_k. What is the (unconditional) distribution of T_k/T_n ?

10. Prove that for $0 < t^* < t$, the conditional distribution of $N(t^*)$, given $N(t)$, is binomial $B(N(t), t^*/t)$, (provided $N(t) \geq 1$).

11. For a homogeneous Poisson process

   (a) Find the coefficient of correlation between $N(t)$ and $N(t + \tau)$ when $t, \tau > 0$.

   (b) Find the coefficient of correlation between T_r and T_s for integers $s > r$.

12. Compute the joint p.d.f. of $(T_1, T_2, \ldots, T_n)$.

13. Prove that the conditional joint distribution of $(T_1, T_2, \ldots, T_{n-1})$, given T_n, is the same as the order statistics of a sample of size $n - 1$ from the uniform distribution over $(0, T_n]$.

14. Do Problem 10 of Section 1.1 and Problem 8 using Equation (1.44). Similarly, do Problems 9 and 10 using the result in Problem 13.

Problems for Section 1.3

1. Suppose that the occurrence over time of failures in some system (the clock is stopped during the repair-time) can be described by a homogeneous Poisson process with some unknown constant failure rate λ. Suppose that the process is observed for some preassigned length of time t and that $N(t)$ is the observed number of failures.

   (a) Show that $N(t)/t$ is UMVUE of λ.

   (b) Verify that the variance of $N(t)/t \to 0$ as $t \to \infty$.

2. Let $0 < t^* < t$. Find the UMVUE for $P\{N(t^*) = k\}$, $0 \leq k \leq N(t)$, on the basis of $N(t)$.
   Hint: Start with a trivial estimator and improve it by the Rao-Blackwell method.

3. A homogeneous Poisson process is observed for 1,000 hours and one sees two occurrences. Using Problem 2, estimate:

   (a) The probability of not observing any occurrences in the first 200 hours.

   (b) The probability of observing exactly one occurrence in the first 200 hours.

4. Suppose we observe a homogeneous Poisson process until the nth failure and record the failure-times $T_1, T_2, \ldots, T_n$. Find a sufficient statistic for the failure-rate λ. Is it complete?

5. Verify that $\hat{\theta} = T_n/n$ is an MLE and unbiased estimator of $\theta = 1/\lambda$, the mean time between occurrences. Compute Var $\hat{\theta}$ and verify that Var $\hat{\theta} \to 0$ as $n \to \infty$.

6. Show that $\lambda^* = (n-1)/T_n$, $n \geq 2$, is the UMVUE of the occurrence rate λ; find its MLE and compute Var(λ^*).

7. Find the UMVUE for $P_k(t^*) = e^{-\lambda t^*}(\lambda t^*)^k/k!$, $k = 0, 1, 2, \ldots, n-1$ on the basis of $T_1, T_2, \ldots, T_n$.

   Hint: Start with $\mathbf{1}\{N(t^*) = k\}$ and improve it by the Rao-Blackwell method.

8. A homogeneous Poisson process is observed for a fixed length of time $t = 10$. Seven occurrences are recorded.

   (a) Compute the conditional probability that the second occurrence takes place after time $t^* = 5$.

   (b) Estimate the probability that the second occurrence takes place after time $t^* = 5$.

   (c) Is there a difference between your answers to (a) and (b)?

9. Suppose that a homogeneous Poisson process is observed until the seventh occurrence is recorded. Suppose that this is at time $t_7 = 10$. Estimate the probability that the second occurrence takes place after time $t^* = 5$.

10. Suppose that a homogeneous Poisson process with event rate λ is observed for a length of time $\min(T_r, t_0)$, where $r > 1$ and t_0 are preassigned. Define $Z = \mathbf{1}\{T_r > t_0\}$ and show that

    (a) The pair $(Z \cdot N(t_0), (1-Z) \cdot T_r)$ is sufficient.

    (b)
    $$\hat{\lambda} = Z \cdot \frac{N(t_0)}{t_0} + (1-Z) \cdot \frac{r-1}{T_r}$$

    is the UMVUE of λ.

    (c) Find the MLE for λ.

    (d) For $k = 0, 1, \ldots, r-1$ and $0 < t^* < t_0$, let $W = \mathbf{1}\{N(t_0) \geq k\}$ and $V = \mathbf{1}\{t^* < T_r\}$. Then
    $$\hat{P}_k(t^*) = Z \cdot W \cdot A + (1-Z) \cdot V \cdot B,$$

    is the UMVUE of $P_k(t^*)$. Here,
    $$A = \binom{N(t_0)}{k}\left(\frac{t^*}{t_0}\right)^k\left(1 - \frac{t^*}{t_0}\right)^{N(t_0)-k}$$

    and
    $$B = \binom{r-1}{k}\left(\frac{t^*}{T_r}\right)^k\left(1 - \frac{t^*}{T_r}\right)^{r-1-k}.$$

    Comment. Note that if $Z = 1$, then $\hat{\lambda}$ is the same as in Problem 1 and if $Z = 0$, then $\hat{\lambda}$ is the same as in Problem 6 (except for notation).

Problems for Section 1.4

1. Fill in the steps to show that $f_k(t)$, the p.d.f. of $T_k = \sum_{i=1}^{k} X_i$, where the X_i are identically distributed independent random variables with common p.d.f. $\lambda e^{-\lambda x}$, $x \geq 0$, is given by

$$\frac{\lambda(\lambda t)^{k-1} e^{-\lambda t}}{(k-1)!}, \qquad t \geq 0.$$

2. Suppose that successive interarrival times are mutually independent, with common p.d.f. $\lambda(\lambda t)^{r-1} e^{-\lambda t}/(r-1)!$, $t \geq 0$ and r a positive integer. Find formulae for $P(N(t) \geq k)$ and $P(N(t) = k)$.

3. Consider a population of items, whose life lengths are exponentially distributed with mean θ. Suppose n of these items are placed on life-test at time $t = 0$. As soon as an item fails, it is replaced instantaneously by another item from the population. Suppose that the first r failure-times (with time measured from $t = 0$, the start of the life-test) occur at times $T_{1,n} \leq T_{2,n} \leq \cdots \leq T_{r,n}$.

   (a) Show that $\{T_{1,n}, T_{2,n} - T_{1,n}, \ldots ; T_{r,n} - T_{r-1,n}\}$, the waiting times between successive failures, are i.i.d. random variables, exponentially distributed with mean θ/n.

   (b) Show that $T_{r,n} \sim \Gamma(r, n/\theta)$.

   (c) Conclude that $E\, T_{r,n} = r\theta/n$ and $\mathrm{Var}\, T_{r,n} = r\theta^2/n^2$.

   (d) Show that the UMVUE of $\lambda = 1/\theta$ is given by $\hat{\lambda} = (r-1)/(nT_{r,n})$.

   (e) Show that the UMVUE of θ is given by $\hat{\theta} = nT_{r,n}/r$.

   (f) Show that the UMVUE of $e^{-\lambda t^*}$ is given by

$$\mathbf{1}\{T_{r,n} > t^*/n\} \cdot \left(1 - \frac{t^*}{nT_{r,n}}\right)^{r-1}.$$

Comment. In this problem, we have a life-test with replacement. Placing n items on test, where the life-time of each item is exponentially distributed with mean θ, and replacing items as they fail by other items whose length of life follows the same p.d.f. is equivalent to observing a superposition of n Poisson processes, each having occurrence rate $\lambda = 1/\theta$. The combined process is also Poisson (see Chapter 2) and has occurrence-rate $n\lambda = n/\theta$.

4. (Continuation of Problem 3.) Suppose that instead of stopping the life-test at the rth failure, life-testing is stopped at a preassigned time t_0. Let K be the observed number of failures in the interval $(0, t_0]$.

   (a) Show that K is a Poisson random variable with parameter nt_0/θ.

   (b) Show that $\hat{\lambda} = K/(nt_0)$ is an unbiased estimator of $\lambda = 1/\theta$.

   (c) Show that if $n\lambda t_0$ is "large" (> 4), then $\hat{\theta} = nt_0/(K+1)$ is a "nearly" unbiased estimator of θ.

(d) Show that $(1 - t^*/(nt_0))^K$ is an unbiased estimator of $e^{-\lambda t^*}$ for any $t^* < nt_0$.

5. (Continuation of Problems 3 and 4.) Suppose that the life-test is stopped at $\min(T_{r,n}, t_0)$.

(a) Show that

$$\hat{\lambda} = \mathbf{1}\{t_0 < T_{r,n}\} \cdot \frac{K}{nt_0} + \mathbf{1}\{t_0 \geq T_{r,n}\} \cdot \frac{r-1}{nT_{r,n}}$$

is an unbiased estimator of λ.

(b) Show that

$$\hat{\theta} = \mathbf{1}\{t_0 < T_{r,n}\} \cdot \frac{nt_0}{K+1} + \mathbf{1}\{t_0 \geq T_{r,n}\} \cdot \frac{nT_{r,n}}{r}$$

is "nearly" unbiased if $n\lambda t_0$ is "large" (> 4).

(c) Show that an unbiased estimator of $e^{-\lambda t^*}$ for $t^* < nt_0$ is provided by

$$\mathbf{1}\{t_0 < T_{r,n}\} \left(1 - \frac{t^*}{nt_0}\right)^K + \mathbf{1}\{t^*/n < T_{r,n} \leq t_0\} \left(1 - \frac{t^*}{nT_{r,n}}\right)^{r-1}.$$

6. An item having exponential life-time with mean $\theta = 1/\lambda$ is placed on life-test. Testing on this item stops either at some preassigned time t_0 or at its time to failure X, whichever comes first. Suppose that item after item is tested in this way until the rth failure occurs. Let N_r (which is a random variable) be the number of items tested. Let $X_1, X_2, \ldots, X_r < t_0$ be the failure-times of the $r(\leq N_r)$ items which failed before time t_0. Define the total time until the rth failure occurs as $T_r = \sum_{i=1}^{r} X_i + (N_r - r)t_0$, the so-called *total time on test*.

(a) Show that $T_r \sim \Gamma(r, \lambda)$.

(b) Conclude that $(r-1)/T_r$ is an unbiased estimator of λ.

(c) Conclude that T_r/r is an unbiased estimator of $\theta = 1/\lambda$.

(d) Show that an unbiased estimator of $P\{T > t^*\} = e^{-\lambda t^*}$, the probability of an item not failing in $(0, t^*]$, is provided by

$$\mathbf{1}\{T_r > t^*\} \left(1 - \frac{t^*}{T_r}\right)^{r-1}.$$

Hint: Verify that the waiting-time until the rth failure occurs has the same distribution as the time to the rth occurrence in a homogeneous Poisson process with rate $\lambda = 1/\theta$.

7. Suppose n items, whose life-times are exponentially distributed with mean $\theta = 1/\lambda$, are placed on test at time $t = 0$. Items that fail are *not* replaced. Suppose that the life-test is terminated when the rth ($r \leq n$) failure occurs. Let $T_{(0)} := 0 < T_{(1)} \leq T_{(2)} \leq \cdots \leq T_{(r)}$ be the first r observed failure times.

(a) Let $Y_j = (n - j + 1)(T_j - T_{j-1})$, $1 \leq j \leq r$. Show that the Y_j are i.i.d. random variables, $Y_j \sim \exp(\lambda)$.

(b) Show that

$$\text{Var}\left(T_{(r)}\right) = \theta^2 \sum_{j=1}^{r} \left(\frac{1}{n - j + 1}\right)^2 .$$

(c) Define the total life on test until the rth failure as

$$T_{r,n} = T_{(1)} + T_{(2)} + \cdots + T_{(r-1)} + (n - r + 1)T_{(r)} .$$

Verify that $\hat{\theta} = T_{r,n}/r$ is the UMVUE of θ.

(d) Show that $\hat{\lambda} = (r - 1)/T_{r,n}$ is the UMVUE of $\lambda = 1/\theta$.

(e) Show that the UMVUE of $e^{-\lambda t^*}$ is provided by

$$\mathbf{1}\{t^* < T_{r,n}\} \cdot \left(1 - \frac{t^*}{T_{r,n}}\right)^{r-1} .$$

Comment. In Problem 7, we have life-testing without replacement. It is useful to think of an equivalent Poisson process with occurrence rate $\lambda = 1/\theta$ associated with each test item. What we are observing is the superposition of these Poisson processes. At time $t = 0$, we start observing a Poisson process with occurrence rate n/θ. When the first failure occurs at $T_{(1)}$, one of the component Poisson processes is "extinguished" and the *interfailure* time $T_{(2)} - T_{(1)}$, is generated by a Poisson process with occurrence rate $(n-1)/\theta$. More generally, the interfailure time $T_{(j)} - T_{(j-1)}$ is generated by a Poisson process with occurrence rate $(n - j + 1)/\theta$, corresponding to the $(n - j + 1)$ processes associated with the unfailed items. It is interesting to note, however, that the total lives between successive failures $Y_1, Y_2, \ldots, Y_n$ can be thought of as being generated by a homogeneous Poisson process with constant rate $\lambda = 1/\theta$. That is, Y_1 is the "time" to the first occurrence in this process, $Y_1 + Y_2$ is the "time" to the second occurrence, and more generally $Y_1 + Y_2 + \cdots + Y_j$ is the "time" to the jth occurrence in a Poisson process having rate $\lambda = 1/\theta$ and $T_{r,n}$ is, of course, $Y_1 + Y_2 + \cdots + Y_r$.

8. (Continuation of Problem 7.) Suppose that the nonreplacement life-test is stopped either at the occurrence of the rth failure or at an accumulated total time on test t_0, whichever occurs first. Let K be the observed number of failures if $\min(T_{r,n}; t_0) = t_0$ (which implies that $0 \leq K \leq r - 1$). Let $Z = \mathbf{1}\{T_{r,n} \leq t_0\}$.

(a) Show that

$$\hat{\theta} = Z \cdot \frac{T_{r,n}}{r} + (1 - Z) \cdot \frac{t_0}{K + 1}$$

is a "nearly" unbiased estimator of θ if t_0/θ is "large" (> 4).

(b) Show that the UMVUE of $\lambda = 1/\theta$ is

$$\hat\lambda = Z \cdot \frac{r-1}{T_{r,n}} + (1-Z) \cdot \frac{K}{t_0}.$$

(c) Define $V = \mathbf{1}\{T_{r,n} > t^*\}$. Show that the UMVUE of $P_0(t^*) = e^{-\lambda t^*}$ for $t^* < t_0$ is

$$\hat{P}_0(t^*) = Z \cdot V \cdot \left(1 - \frac{t^*}{T_{r,n}}\right)^{r-1} + (1-Z) \cdot \left(1 - \frac{t^*}{t_0}\right)^K.$$

Problems for Section 1.5

In this section, the basic assumption is that we observe a time-dependent Poisson process with intensity $\lambda(t)$ and mean measure $\Lambda(t) = \int_0^t \lambda(s)ds$.

1. Let T_1 be the first occurrence-time. Prove that

$$P(T_1 > t+\tau \mid T_1 > t) = e^{-[\Lambda(t+\tau)-\Lambda(t)]}.$$

What does this mean in the special case where $\lambda(t) \equiv \lambda$ (a constant)?

2. Prove Equations (1.50), (1.51), (1.52), (1.56), and (1.57).

3. Show that

$$E(T_k) = \sum_{j=0}^{k-1} \int_0^\infty \frac{[\Lambda(t)]^j}{j!} e^{-\Lambda(t)}dt$$

and further that

$$E(T_{k+1} - T_k) = \int_0^\infty \frac{[\Lambda(t)]^k}{k!} e^{-\Lambda(t)}dt.$$

4. (Continuation of Problem 3.) Evaluate $E(T_k)$ and $E(T_{k+1} - T_k)$ if $\Lambda(t) = (\alpha t)^\beta$ for $k = 1, 2, \ldots$. Verify that $E(T_{k+1} - T_k)$ is an increasing function of k if $\beta < 1$, is a decreasing function of k if $\beta > 1$.

5. Let $T_1, T_2, \ldots, T_j, \ldots$ be the times when the first, second,$\ldots, j^{\text{th}}, \ldots$ events occur. Define the random variables $\Lambda(T_1), \Lambda(T_2), \ldots, \Lambda(T_j), \ldots$. Show that the random variables $\Lambda(T_1), \Lambda(T_2) - \Lambda(T_1), \ldots, \Lambda(T_j) - \Lambda(T_{j-1}), \ldots$ are i.i.d. unit-exponential.

6. Prove that the random variable $2\Lambda(T_j)$ is distributed as $\chi^2(2j)$.

7. Suppose that $\lambda(t)$ increases (decreases) monotonically in t. Show that $\Lambda(t)/t$ will also increase (decrease) monotonically in t.

8. Suppose that successive failure-times of a machine can be considered as being generated by a time-dependent Poisson process with failure rate $\lambda(t) = .01 + .5t^{-\frac{1}{2}}$. What is the probability that no failures will occur in the intervals [25, 30]; [100, 105]; [2,500, 2,505]; [10,000, 10,005]?

9. Suppose that successive failure times of a device can be considered as being generated by a time-dependent Poisson process with failure rate $\lambda(t) = .005 + (.001)e^{.01t}$. What is the probability of no failure occurring in the intervals $[0, 10]$; $[100, 110]$; $[1,000, 1,010]$?

10. Consider a time-dependent Poisson process for which $\Lambda(t)/t$ is monotonically increasing (decreasing). Suppose that $e^{-\Lambda(t^*)} = p^*$. Show that

$$e^{-\Lambda(t)} \overset{\geq}{\underset{(\leq)}{}} (p^*)^{t/t^*}, \qquad \text{for } t \leq t^*$$

and

$$e^{-\Lambda(t)} \overset{\leq}{\underset{(\geq)}{}} (p^*)^{t/t^*}, \qquad \text{for } t \geq t^*.$$

11. A time-dependent Poisson process with event rate $\lambda(t)$ is observed for a fixed length of time t_0. Show that for $0 < t^* < t_0$, the conditional distribution of $N(t^*)$, given $N(t_0)$, is $B(N(t_0), \Lambda(t^*)/\Lambda(t_0))$.

12. (Continuation of Problem 11.) Compute $E[1 - \Lambda(t^*)/\Lambda(t_0)]^{N(t_0)}$.

13. (Continuation of Problem 12.) Suppose that $\Lambda(t)/t$ is monotonically increasing. Show that $(1 - t^*/t_0)^{N(t_0)}$ provides a "conservative" lower bound for $e^{-\Lambda(t^*)}$ in the sense that $e^{-\Lambda(t^*)} \geq E(1 - t^*/t_0)^{N(t_0)}$.

14. (Continuation of Problem 13.) Suppose that $\Lambda(t)/t^\beta$, $\beta \geq 1$ is monotonically increasing. Show that $[1 - (t^*/t_0)^\beta]^{N(t_0)}$ provides a "conservative" lower bound for $e^{-\Lambda(t^*)}$.

15. Suppose that a nonhomogeneous Poisson process with event rate $\lambda(t) = \alpha t$ is observed for a length of time $t_0 = 1,000$. A total of 10 failures are observed during this interval. Estimate the probability of no occurrences in the interval $(0, 100]$. What could we have said about this probability if we had known only that $\lambda(t)$ is monotonically increasing in t?
Hint: Use Problem 12.

Problems for Section 1.6

The basic assumption in this section is that we observe a Bernoulli sequence $\{X_i\}$, with parameter p. We denote by S_k the number of successes in the first k trials and by Y_k the trial on which the kth success occurs.

1. Verify that $\sum_{k=0}^{n} P(S_n = k) = 1$ and that $E(S_n) = np$ and $\text{Var}(S_n) = npq$.

2. Verify that $E(Y_1) = 1/p$ and $\text{Var}(Y_1) = q/p^2$.

3. Verify that $E(Y_k) = k/p$ and $\text{Var}(Y_k) = kq/p^2$.
   Hint: $Y_k = Y_1 + (Y_2 - Y_1) + \cdots + (Y_k - Y_{k-1})$.

4. Verify the equivalence of the events $Y_k > j$ and $S_j \le k - 1$, $Y_1 > j$ and $S_j = 0$ and express this equivalence in words.

5. Prove that $P(Y_1 = j + k \mid Y_1 > j) = pq^{k-1}$, $k = 1, 2, \ldots$. State a continuous time analogue of this result.

6. The random variable $Z_k = Y_k - k$ is the number of failures observed prior to the occurrence of the kth success. Find the probability distribution of Z_k and also find $E(Z_k)$ and $\text{Var}(Z_k)$.

7. Suppose that the probability of a successful rocket launching is $p = .6$.

   (a) Find the probability of at least one success in the next 5 launchings.

   (b) What is the probability of exactly 3 successes?

   (c) What is the probability of at most 3 successes?

   (d) What is the probability of at least 3 successes?

   (e) Find the mean and variance of the number of successes.

8. Suppose that the probability of a successful moon shot is .1.

   (a) What is the probability that the first success occurs on the fifth shot?

   (b) What is the probability that it will take at least 5 shots to produce a success?

   (c) Suppose that the first 5 shots have ended in failure. What is the probability that the first success occurs on the tenth shot given this information?

   (d) Find the expectation and variance of the number of trials to the first success.

9. (Continuation of Problem 8.)

   (a) What is the probability that the second success occurs on the fifth trial?

   (b) What is the probability that the third success occurs on the tenth trial?

   (c) What is the probability that the third success occurs on or before the tenth trial?

10. Suppose we observe a Bernoulli sequence until the nth trial.

    (a) Find a sufficient statistic for p. Is it complete?

    (b) Find MLE and MOM estimators for p.

    (c) Is any of them UMVUE ?

    (d) Verify that $\text{Var}(S_n/n) \to 0$ as $n \to \infty$.

    (e) Prove that there is no unbiased estimator for $1/p$, which is based on S_n.

11. Suppose the Bernoulli sequence is observed until the kth success.

    (a) Find a sufficient statistic. Is it complete?

    (b) Verify from Problem 3 that Y_k/k is an unbiased estimator of $1/p$ and verify that $\text{Var}(Y_k/k) \to 0$ as $k \to \infty$.

12. Verify that $(k-1)/(Y_k - 1)$, $k = 2, 3, \ldots$ is an unbiased estimator of p.

Comment. Note the essential difference between the estimators in Problems 10 and 12. In Problem 10 the experiment consists of making a preassigned number of Bernoulli trials n and basing the estimate of p on the random variable S_n, the number of observed successes in n trials. In Problem 12 the experiment consists of continuing to make Bernoulli trials until one has observed a preassigned number of successes, k. In this case, the estimator is based on the random variable Y_k, the trial on which the kth success occurs.

13. Suppose that a sequence of satellite firings can be considered as successive Bernoulli trials with a common probability, p, of success on each trial. Suppose that five successes are observed in the first ten firings.

    (a) Give an unbiased estimate for p.

    (b) What is the conditional probability that exactly two successes occurred in the first five firings?

    (c) What is the conditional probability that the second success occurred on the fifth trial?

    (d) What is the conditional probability that the second success occurred after the fifth trial?

14. As in Problem 13, we consider a sequence of satellite firings with common probability, p, of success on each trial. Testing is continued until the fifth success occurs. Suppose that this happens on the tenth trial. Answer the same questions as in Problem 13.

15. Suppose that we make a sequence of Bernoulli trials where the probability of success on each trial is p. We stop making trials at $\min(Y_k, n)$, where k and n are preassigned integers with $k \le n$.

    (a) What is a sufficient statistic?

    (b) Show that

$$\mathbf{1}\{Y_k > n\} \cdot \frac{S_n}{n} + \mathbf{1}\{Y_k \le n\} \cdot \frac{k-1}{Y_k - 1}$$

    is UMVUE of p.

16. Prove that for $1 \le n^* \le n$, the conditional distribution of S_{n^*}, given S_n, is *hypergeometric*. That is,

$$P(S_{n^*} = j \mid S_n = k) = \frac{\binom{n^*}{j}\binom{n-n^*}{k-j}}{\binom{n}{k}},$$

    where $\max(0, k - n + n^*) \le j \le \min(k, n^*)$.   (1.63)

    Verify that $E(S_{n^*} \mid S_n = k) = kn^*/n$.

17. (Continuation of Problem 16.) Suppose that we wish to estimate p^{n^*}, the

probability that no failures occur in $n^* < n$ trials. Verify that

$$1\{S_n \geq n^*\} \cdot \frac{\binom{n-n^*}{S_n-n^*}}{\binom{n}{S_n}}$$

provides an unbiased estimator of p^{n^*} (in fact it is UMVUE).

18. (Continuation of Problem 17.) Verify that

$$1\{n^* - j \leq S_n \leq n - j\} \cdot \frac{\binom{n^*}{n^*-j}\binom{n-n^*}{S_n-n^*+j}}{\binom{n}{S_n}}$$

provides an unbiased estimator of $P\{S_{n^*} = j\}$.

19. Items are being produced which have probability p of being good and probability $q = 1 - p$ of being defective. One hundred of these items are produced. Of these, 97 are good and 3 are defective. Use Problems 17 and 18 to estimate the probability that there are no defective items in a sample of ten items; that there is at most one defective in a sample of ten items; that there are at most two defectives in a sample of ten items.

20. Prove that

$$P(Y_s = j \mid S_n = k) = \frac{\binom{j-1}{s-1}\binom{n-j}{k-s}}{\binom{n}{k}},$$

where $1 \leq s \leq k$, $s \leq j \leq n - k + s$.

21. Prove that

$$P(Y_s = j \mid Y_r = k) = \frac{\binom{j-1}{s-1}\binom{k-j-1}{r-s-1}}{\binom{k-1}{r-1}},$$

where $1 \leq s < r \leq k$; $s \leq j \leq k - r + s$. In particular, verify that $P(Y_1 = j \mid Y_2 = k) = 1/(k-1)$, $1 \leq j \leq k - 1$.

22. A sequence of Bernoulli trials with probability of success p on each trial is conducted until the kth success occurs, where $k \geq 1$ is a preassigned integer. Suppose that the trial on which the kth success occurs is $Y_k = n$. Let S_{n^*} be the number of successes in the first $n^* < n$ trials. Prove that

$$P(S_{n^*} = j \mid Y_k = n) = \frac{\binom{n^*}{j}\binom{n-n^*-1}{k-j-1}}{\binom{n-1}{k-1}},$$

where $\max(0, k - n + n^*) \leq j \leq \min(k - 1, n^*)$.

23. (Continuation of Problem 22.) Suppose that we wish to estimate p^{n^*}, the probability that no failures occur in n^* trials. Verify that for $n^* \leq k - 1$,

$$\frac{\binom{Y_k-n^*-1}{k-n^*-1}}{\binom{Y_k-1}{k-1}}$$

is an unbiased estimator of p^{n^*}.

24. Items are being produced which have probability p of being good and probability $q = 1 - p$ of being defective. Items are tested one at a time until 100 good items are found. The 100th good item is obtained on the 105th item tested. Estimate the probability that ten items in a row are all good.

25. Let now Z_k be the trial on which the kth *failure* occurs and let $F_n = n - S_n$ be the number of failures among the first n trials. Suppose we stop at Z_k, where k is preassigned. Let $n^* < n$. Show that

$$P(S_{n^*} = j \mid Y_k = n) = P(F_{n^*} = j \mid Z_k = n).$$

26. (Continuation of Problem 25). Verify that

$$\mathbf{1}\{Z_k \geq n^* + k\} \cdot \frac{\binom{Z_k - n^* - 1}{k-1}}{\binom{Z_k - 1}{k-1}}$$

provides an unbiased estimator of p^{n^*}, the probability of getting n^* successes in a row.

27. A sequence of Bernoulli trials is conducted until the fifth failure occurs. Suppose that this happens on the 105th trial. Estimate the probability that ten trials in a row are all successful.

CHAPTER 2

Statistical life length distributions

2.1 Stochastic life length models

In this chapter we present some stochastic models for length of life. In particular, we wish to consider in some detail why one may expect the exponential, gamma, Weibull, and certain other distributions to occur. We shall see that the considerations in the first chapter concerning the Poisson process play a fundamental role. In large part, this chapter is based on Epstein [20].

2.1.1 Constant risk parameters

We begin by describing the following very simple model. Imagine a situation where a device is subject to an environment or risk E, which can be described as a random process in which "dangerous" peaks occur with Poisson rate λ. It is assumed that only these peaks can affect the device in the sense that the device will fail if a peak occurs and will not fail otherwise. If this is the situation, then the time to failure distribution must be exponential. More precisely, if T denotes the life time of the device, then

$$P(T > t) = P[\text{no shock in the interval } (0, t]] = e^{-\lambda t}. \qquad (2.1)$$

Hence the c.d.f. of T is

$$F(t) = P(T \le t) = 1 - e^{-\lambda t}, \qquad t \ge 0 \qquad (2.2)$$

and the p.d.f. is

$$f(t) = \lambda e^{-\lambda t}, \qquad t \ge 0. \qquad (2.3)$$

It is not necessary that we have an all-or-none situation in order that the exponential distribution arises. Suppose, as before, that dangerous peaks in the stochastic process occur with rate λ and that the conditional probability that the device fails (survives), given that a peak has occurred, is $p(q = 1 - p)$. It is also assumed that the device has no memory. That is, if it has survived $(k - 1)$

peaks, then the conditional probability of failing because of the kth peak is p. Then it is clear that the event, "device does not fail in $(0, t]$", is composed of the mutually exclusive events:

"no peaks occur in $(0, t]$";

"one peak occurs in $(0, t]$ and the device does not fail";

"two peaks occur in $(0, t]$ and the device does not fail";

etc.

Hence, the probability that the device has a life time T exceeding t, is given by

$$
\begin{aligned}
P(T > t) &= e^{-\lambda t} + q\lambda t e^{-\lambda t} + q^2 \frac{(\lambda t)^2}{2!} e^{-\lambda t} + \cdots + q^k \frac{(\lambda t)^k}{k!} e^{-\lambda t} + \cdots \\
&= e^{-\lambda t} \left[1 + q\lambda t + \frac{(q\lambda t)^2}{2!} + \cdots \right] \\
&= e^{-\lambda t + \lambda q t} = e^{-\lambda(1-q)t} = e^{-\lambda p t}.
\end{aligned}
\tag{2.4}
$$

Thus, the c.d.f. of T is given by

$$
G(t) = P(T \le t) = 1 - e^{-\lambda p t}, \qquad t \ge 0
\tag{2.5}
$$

and the p.d.f. $g(t)$ is

$$
g(t) = \lambda p e^{-\lambda p t}, \qquad t \ge 0.
\tag{2.6}
$$

Again, we have an exponential distribution. The effective failure rate is now λp and the mean time to failure is given by $1/(\lambda p)$.

Carrying this model further, suppose that a device is exposed simultaneously to k competing risks (or environments) $E_1, E_2, \ldots, E_k$. Suppose further that λ_i is the rate of occurrence of dangerous peaks in environment E_i and that the conditional probability of the device failing given that this peak has occurred is p_i. From Equation (2.4) it follows that the probability that a device survives environment E_i for a length of time t is given by $e^{-\lambda_i p_i t}$. If the environments compete independently as causes for failure of the device, then from Equation (2.4) and the assumption of independence, the time T until a failure occurs is a random variable such that

$$
P(T > t) = \prod_{i=1}^{k} \exp[-\lambda_i p_i t] = \exp\left[-\left(\sum_{i=1}^{k} \lambda_i p_i \right) t \right] = e^{-\Lambda t}, \qquad t \ge 0
\tag{2.7}
$$

where $\Lambda = \sum_{i=1}^{k} \lambda_i p_i$.

The c.d.f. of T is given by

$$
F(t) = P(T \le t) = 1 - e^{-\Lambda t}, \qquad t \ge 0.
\tag{2.8}
$$

Thus, we are again led to an exponential life distribution.

2.1.2 Time-dependent risk parameters

In the situation described by the c.d.f. (2.5) and p.d.f. (2.6), p, the conditional probability that a failure occurs, given that a peak has occurred, is independent of t. Let us now assume that the conditional probability of a failure, given a peak at time t, is given by a nonnegative integrable function $p(t)$, $0 \leq p(t) \leq 1$, for all t. It can be shown in this case that the probability that the time to failure T exceeds t is given by

$$P(T > t) = \exp\left[-\lambda \int_0^t p(\tau)d\tau\right].$$ (2.9)

Hence, the c.d.f. of T is given by

$$F(t) = P(T \leq t) = 1 - \exp[-\lambda P(t)], \qquad t \geq 0$$ (2.10)

and the p.d.f. of T is given by

$$f(t) = \lambda p(t) \exp[-\lambda P(t)], \qquad t \geq 0,$$ (2.11)

where $P(t) = \int_0^t p(\tau)d\tau$.

To prove Equation (2.9) we argue as follows. Let $N(t)$ be the number of peaks in $(0, t]$. Looking at Equation (2.4), the kth term in the first line is $P\{B \cap A_k\} = P\{A_k\}P\{B|A_k\}$, where $A_k = \{N(t) = k\}$ and $B = \{$the device survives$\}$. But given A_k, the peak-times S_i are distributed as k i.i.d. uniform random variables in $(0, t]$. Hence, the constant q^k is replaced by the conditional mean of the product $q(S_1)q(S_2)\ldots q(S_k)$. Since

$$Eq(S_i) = \frac{1}{t}\int_0^t (1 - p(s))ds = \frac{t - P(t)}{t},$$

we have

$$P\{B \cap A_k\} = \frac{e^{-\lambda t}(\lambda t)^k}{k!} \cdot \frac{(t - P(t))^k}{t^k} = \frac{e^{-\lambda t}(\lambda(t - P(t)))^k}{k!}.$$

Summing over all $k \geq 0$ gives Equation (2.9).

The distribution of T is given by the one-parameter exponential distribution if and only if $p(t) \equiv p$, $0 < p \leq 1$, for all $t \geq 0$.

Examples:

(1) Suppose that

$$\begin{aligned} p(t) &= 0, & 0 \leq t < A \\ &= 1, & t \geq A. \end{aligned}$$

Then it is readily verified that the c.d.f. and p.d.f. of T are given by

$$\begin{aligned} F(t) &= 0, & t < A \\ &= 1 - e^{-\lambda(t-A)}, & t \geq A \end{aligned}$$

and
$$f(t) = \quad 0, \qquad\qquad t < A$$
$$= \quad \lambda e^{-\lambda(t-A)}, \qquad t \geq A.$$

The p.d.f. $f(t)$ is known as the two-parameter (or shifted) exponential distribution.

(2) Suppose that
$$p(t) = \quad \left(\frac{t}{t_0}\right)^{\alpha}, \quad 0 \leq t < t_0$$
$$= \quad 1, \qquad t \geq t_0.$$

In this case, the c.d.f. of T is given by
$$F(t) = \quad 1 - \exp\left[-\frac{\lambda t^{\alpha+1}}{(\alpha+1)t_0^{\alpha}}\right], \qquad\qquad 0 \leq t < t_0$$
$$= \quad 1 - \exp\left[-\lambda\left(t - \frac{\alpha}{\alpha+1}t_0\right)\right], \qquad t \geq t_0.$$

Note, this distribution is continuous; in the range $0 \leq t < t_0$, it is a Weibull distribution, then, a shifted exponential.

In the model just described, we can think of peaks in the environment being generated by a homogeneous Poisson process with constant event rate λ per unit time and with time-dependent conditional probability $p(t)$ of a device failing, given that a peak has occurred. An alternative model is to assume that peaks in the environment are generated by a time-dependent Poisson process with rate $\lambda(t)$, with conditional probability of failure, given a peak, equal to p. For the alternative model, the analogues of Equations (2.10) and (2.11) are, respectively,
$$F(t) = P(T \leq t) = 1 - \exp[-p\Lambda(t)], \qquad t \geq 0 \qquad (2.12)$$
and
$$f(t) = p\lambda(t) \exp{-[p\Lambda(t)]}, \qquad t \geq 0, \qquad (2.13)$$
where $\Lambda(t) = \int_0^t \lambda(\tau)d\tau$.

2.1.3 Generalizations

A generalization in still another direction is to assume that although each device is exposed to one and only one environment (risk), there are k possible environments (risks), e.g., different weather conditions, $E_1, E_2, \ldots, E_k$, which can occur with respective probabilities $c_1, c_2, \ldots, c_k$ ($c_j \geq 0$, $\sum_{j=1}^{k} c_j = 1$). Furthermore, with each environment E_j there is associated a λ_j (the rate of occurrence of dangerous peaks in environment E_j) and devices exposed to environment E_j will fail with probability p_j, given that an environmental peak has occurred. Using the theorem of total probability, it follows that the probability that the time to failure, T, of a device exceeds t is given by

$$P(T > t) = \sum_{j=1}^{k} c_j \exp[-\lambda_j p_j t], \qquad t \geq 0. \tag{2.14}$$

The associated p.d.f. $f(t)$ is given by

$$f(t) = \sum_{j=1}^{k} c_j \nu_j \exp[-\nu_j t], \qquad t \geq 0 \tag{2.15}$$

where $\nu_j = \lambda_j p_j$. This is called a *convex combination* or a *mixture* of exponential distributions, with weights c_j. The model can be regarded as a *Bayesian model*, where the weights reflect one's belief about the weather conditions under which the device is supposed to operate. The p.d.f. (2.15) arises frequently in biostatistical situations. There has been considerable research on the problem of estimating the parameters of this model.

A continuous analogue of Equation (2.15) is given by

$$f(t) = \int_0^\infty \nu e^{-\nu t} dG(\nu), \qquad t \geq 0, \tag{2.16}$$

where G is the prior distribution of the failure-rate. Chapter 6 of Crowder et al. [13] deals with Bayesian statistical analysis of reliability data. Another practical treatment of Bayesian reliability estimation and prediction is given by Zacks [56].

Examples: Suppose that

$$g(\nu) = \frac{dG(\nu)}{d\nu} = \frac{\nu^r A^{r+1} e^{-A\nu}}{r!}, \qquad \nu \geq 0, \tag{2.17}$$

i.e., G is the $\Gamma(r+1, A)$ (gamma) distribution. Equation (2.16) then becomes

$$f(t) = \int_0^\infty \frac{\nu^{r+1} A^{r+1} e^{-\nu(A+t)}}{r!} d\nu, \qquad t \geq 0. \tag{2.18}$$

Carrying out the integration, we get

$$f(t) = \frac{(r+1)A^{r+1}}{(A+t)^{r+2}}, \qquad t \geq 0, \tag{2.19}$$

which is a *Pareto* p.d.f. (defined by $f(x) = \alpha c^\alpha (c+x)^{-\alpha-1}$, $x \geq 0$, $c > 0$) with shape $\alpha = r + 1$, and scale $c = A$. Like the one parameter exponential distribution, $f(t)$ decreases monotonically to zero and has its mode at $t = 0$. It is interesting to note that in the special case $A = r + 1$, the life time T is distributed according to an $F(2,\ 2r+2)$ distribution.

An alternative model for length of life resulting in a mixture of exponential distributions with the p.d.f. $\sum_{j=1}^{k} c_j \lambda_j e^{-\lambda_j t}$, $t \geq 0$, can be obtained as follows. Let us assume that the devices fall into k types with probability of occurrence for Type j equal to c_j, where $\sum_{j=1}^{k} c_j = 1$. Furthermore, suppose that a device of Type j can fail in one and only one way, which we call a Type j failure. The p.d.f. of a Type j failure is assumed to be $\lambda_j e^{-\lambda_j t}$. Using the Law of Total Probability, we obtain the desired result. For a discussion of this model for $k = 2$, see Chapter 10 of Cox [9].

Another generalization of the shock model is to assume that a device subject to an environment E will fail when exactly $k \geq 1$ shocks occur and not before. If shocks are generated by a Poisson process with rate λ, then it follows from the results of Chapter 1 that the p.d.f. and c.d.f. of T_k are given respectively by

$$f_k(t) = \frac{\lambda^k t^{k-1} e^{-\lambda t}}{(k-1)!}, \qquad t \geq 0 \qquad (2.20)$$

and

$$F_k(t) = 1 - \sum_{j=0}^{k-1} \frac{e^{-\lambda t} (\lambda t)^j}{j!}, \qquad t \geq 0. \qquad (2.21)$$

Formulae (2.5), (2.6), (2.10), (2.11), and (2.16) can be suitably generalized to the case where the life time is given by T_k and not by T_1. Thus, for example, the analogues of (2.10) and (2.11) become

$$F_k(t) = P(T_k \leq t) = 1 - \sum_{j=0}^{k-1} e^{-\lambda P(t)} \frac{[\lambda P(t)]^j}{j!}, \qquad t \geq 0 \qquad (2.22)$$

and

$$f_k(t) = \lambda p(t) \frac{[\lambda P(t)]^{k-1} e^{-\lambda P(t)}}{(k-1)!}, \qquad t \geq 0. \qquad (2.23)$$

To sum up, we have seen in this section that the simple assumption that failure of a device is associated with the occurrences of peaks in a Poisson process leads to a wide variety of distributions. In particular, this model generates the exponential (one and two parameter), gamma, mixed exponential (its p.d.f. is a convex combination of two or more exponential p.d.f.s), Weibull and Pareto distributions.

It should be noted that none of the failure models considered lead to a normal distribution of length of life. How then does one explain the fact that some observed life time distributions appear to be normal? One of the ways in which the normal distribution can arise is very clear. Suppose that k, the number of shocks required for failure, is large. Then the p.d.f. (2.20) (a gamma distribution) is, for large k, approximately normal with mean k/λ and variance k/λ^2 (this is an immediate consequence of the central limit theorem). For example, we might think of an object undergoing a certain amount of wear (possibly constant) each time that a shock occurs and we can assume further that when the total wear exceeds some preassigned value, failure will occur. If the number of steps required to attain the critical amount of wear is large, then the time to failure is the sum of a large number of independent random variables and will be approximately normally distributed. For a discussion of this type of model, see Chapters 8 and 10 of Cox [9].

Another possibility is that failure occurs after an essential substance has been used up. In this case, the time to failure might be proportional to the amount of this substance in the particular specimens being tested. If the amount of the substance varies from specimen to specimen, according to a normal distribution, then one would get a normal distribution of life time.

It is interesting to note that the failure distribution (2.23) is almost identical with the Birnbaum-Saunders [7] model. This is a statistical model for life-length of structures under dynamic loading. For a detailed discussion, see Epstein [20].

2.2 Models based on the hazard rate

In this section, we discuss a failure mode, which involves the conditional failure or *hazard rate*. The *hazard rate*, also called *hazard function* is precisely the same as the *force of mortality* introduced by actuaries many years ago. More precisely, let the length of life T of some device be a random variable with p.d.f. $f(t)$ and c.d.f. $F(t)$, respectively. We wish to compute the probability that the device will fail in the time interval $(t, t+\Delta t]$ given that it has survived for a length of time t, i.e., we wish to compute $P(t < T \leq t + \Delta t \mid T > t)$. Using the definition of conditional probability, it follows that

$$P(t < T \leq t + \Delta t \mid T > t) = P(t < T \leq t + \Delta t, T > t)/P(T > t). \tag{2.24}$$

But the event $\{t < T \leq t + \Delta t\}$ is included in the event $\{T > t\}$ (in other words failure in the interval $(t, t + \Delta t]$ is possible only if the device has survived up to time t). Hence Equation (2.24) becomes

$$P(t < T \leq t + \Delta t \mid T > t) = \frac{f(t)\Delta t + o(\Delta t)}{1 - F(t)}. \tag{2.25}$$

The conditional probability of failing in $(t, t + \Delta t]$ given survival up to time t is converted into a (conditional) failure rate or hazard rate by dividing by Δt

and letting $\Delta t \to 0$. Thus, the hazard rate, $h(t)$, associated with a p.d.f. $f(t)$ is

$$h(t) = \lim_{\Delta t \to 0} \frac{P(t < T \leq t + \Delta t \mid T > t)}{\Delta t} = \frac{f(t)}{1 - F(t)} . \qquad (2.26)$$

It is just as easy to find $f(t)$ and $F(t)$ from $h(t)$ as it was to find $h(t)$ from $f(t)$. This is readily done since Equation (2.26) can also be written as

$$\frac{d}{dt} \log(1 - F(t)) = -h(t) . \qquad (2.27)$$

Integrating both sides of Equation (2.27) over $(0, t]$ and recalling that $F(0) = 0$, we obtain

$$F(t) = 1 - e^{-H(t)}, \qquad t \geq 0 , \qquad (2.28)$$

where $H(t) = \int_0^t h(\tau)d\tau$ is the *integrated* or *cumulative hazard function*.

The p.d.f. $f(t)$ is obtained by differentiating Equation (2.28) which gives

$$f(t) = h(t) e^{-H(t)}, \qquad t \geq 0 . \qquad (2.29)$$

The *reliability* $R(t)$ of a device over the interval $(0, t]$ is defined as the probability that the device does not fail over this time interval. Hence, it follows from Equation (2.28) that

$$R(t) = P(T > t) = 1 - F(t) = e^{-H(t)}, \qquad t \geq 0 . \qquad (2.30)$$

It should be noted that the c.d.f.s given by Equation (2.28) are direct generalizations of the simple exponential distribution. Hence, it is possible to change the time scale in such a way that results obtained under a purely exponential assumption are valid for a more general situation. More precisely, if the random variable T represents the time to failure when the underlying hazard function is $h(t)$, then the random variable $U = H(T) = \int_0^T h(\tau)d\tau$ is distributed with c.d.f. $1 - e^{-u}$, $u \geq 0$ and p.d.f. e^{-u}, $u \geq 0$. Consequently, many of the theorems, formulae, estimation procedures, etc., given in life-test theory when the underlying distribution is exponential, become valid in a much more general situation, if one replaces the failure times T_i, by generalized times $U_i = H(T_i)$.

Examples:

(1) Suppose that $h(t) \equiv 1/\theta$, $t \geq 0$. Then Equation (2.28) becomes

$$F(t) = 1 - e^{-t/\theta}, \qquad t \geq 0$$

and Equation (2.29) becomes

$$f(t) = \frac{1}{\theta} e^{-t/\theta}, \qquad t \geq 0 .$$

Thus, we see that the statement, "the device has a constant hazard rate" implies that the underlying life distribution is exponential. Conversely, if the underlying life distribution is exponential, the hazard rate is constant.

(2) Suppose that

$$h(t) = 0, \qquad 0 \le t < A$$
$$= \tfrac{1}{\theta}, \qquad t \ge A.$$

In this case, we get the two parameter exponential distribution and Equations (2.28) and (2.29) become

$$F(t) = 0, \qquad t < A$$
$$= 1 - \exp\left[-\left(\frac{t-A}{\theta}\right)\right], \qquad t \ge A$$

$$f(t) = 0, \qquad t < A$$
$$= \frac{1}{\theta}\exp\left[-\left(\frac{t-A}{\theta}\right)\right], \qquad t \ge A.$$

(3) Suppose that $h(t) = \alpha t^{\alpha-1}/\theta^{\alpha}$, where $\alpha > 0$. In this case Equations (2.28) and (2.29) become

$$F(t) = 1 - \exp\left[-\left(\frac{t}{\theta}\right)^{\alpha}\right], \qquad t \ge 0$$

and

$$f(t) = \frac{\alpha t^{\alpha-1}}{\theta^{\alpha}}\exp\left[-\left(\frac{t}{\theta}\right)^{\alpha}\right], \qquad t \ge 0.$$

This distribution is often called the *Weibull distribution* after the Swedish physicist and engineer who found that this distribution gave reasonable empirical fits to fatigue data. It should more properly be called Gumbel's type three asymptotic distribution of smallest values (see [21] and [31]). Another name for the Weibull distribution is the Rosin-Rammler law, used in fitting the distribution of particle sizes obtained after the physical breakage or grinding of certain materials.

Perhaps the popularity of the Weibull distribution is connected with the fact that its associated hazard function is simply a power of t. The constant α is called the *shape parameter* and the value of α determines whether the hazard rate $h(t) = \alpha\theta^{-\alpha}t^{\alpha-1}$ is decreasing, constant, or increasing. More precisely, $h(t)$ is

$$\begin{array}{lll} \text{decreasing,} & \text{if} & 0 < \alpha < 1 \\ \text{constant,} & \text{if} & \alpha = 1 \\ \text{increasing,} & \text{if} & \alpha > 1. \end{array}$$

(4) Suppose $h(t) = \alpha e^{\beta t}$, $t \geq 0$. Associated with this hazard function are the c.d.f.

$$F(t) = 1 - \exp\left[-\frac{\alpha}{\beta}\left(e^{\beta t} - 1\right)\right], \qquad t \geq 0$$

and p.d.f.

$$f(t) = \alpha e^{\beta t} \exp\left[-\frac{\alpha}{\beta}\left(e^{\beta t} - 1\right)\right], \qquad t \geq 0.$$

This is the Gumbel distribution which is the asymptotic distribution of smallest values. The slightly more general hazard function, $h(t) = \lambda_0 + \alpha e^{\beta t}$, where λ_0, α, β are all positive results in a c.d.f. discovered by the actuary Makeham. Mortality data for adult human populations, times to failure of aircraft engines, etc., can be fitted reasonably well by the appropriate choice of λ_0, α, β.

We have seen the Gumbel and Weibull distributions of smallest values as possible life distributions in examples (3) and (4). This was done by choosing appropriate hazard functions. There are actually good physical reasons for expecting the extreme value distributions, if failure occurs at what might be termed a *weakest link* or weakest spot. Thus, for example, we can think of corrosion as a phenomenon involving the chemical perforation of a surface, covered with many imperfections or pits. The time to failure may then be viewed as follows. Let T_i be the time required to penetrate the surface at the ith pit. Then the time to failure is $\min\{T_i\}$. Suppose, for example, that pit depths follow an exponential distribution and that the time to perforation at a pit is proportional to the thickness of the coating minus the pit depth. Then, extreme value theory would lead one to expect a life time p.d.f. of the form $A\beta \exp(\beta t) \exp(-A e^{\beta t})$. For further details on extreme values in this connection, see [18], [19], [21] and [23]. For the general theory of extreme value theory see de Haan and Ferreira [15], Beirlant et al. [4], or Smith and Weissman [46].

2.2.1 IFR and DFR

Considerable work has been done on life distributions whose associated hazard functions are monotone, i.e., either *increasing failure rate* (IFR) or *decreasing failure rate* (DFR). (By "increasing," we mean "monotonically nondecreasing." By "decreasing," we mean "monotonically nonincreasing." The exponential life distribution $F(t) = 1 - e^{-t/\theta}$, $t \geq 0$ with associated hazard function $h(t) \equiv 1/\theta$ for all t is a boundary case and is simultaneously IFR and DFR. It is intuitively clear that for an item whose life distribution is IFR, a *new* item is *better* than *used* (NBU) in the sense that

$$P(T > t + s \mid T > t) \leq P(T > s),$$

or equivalently

$$R(t + s) \leq R(t)R(s).$$

For DFR distribution, *new is worse* than *used* (NWU) with reverse inequalities. These concepts are treated in Barlow and Proschan [2]. Since details on IFR and DFR distributions can be found in Barlow and Proschan [3], we limit ourselves to giving two results, which are both useful and easily established. For convenience, we treat the IFR case, where $h(t)$ is assumed to be increasing.

Suppose that $F(t)$ is an IFR distribution and that ξ_p is its p–quantile ($F(\xi_p) = p$). For convenience, we are assuming that $F(t)$ is continuous and strictly increasing, so that ξ_p is uniquely determined. We then assert that

$$
\begin{aligned}
R(t) = P(T > t) = 1 - F(t) \quad &\geq \quad (1-p)^{t/\xi_p} \text{ for } t \leq \xi_p \\
&\leq \quad (1-p)^{t/\xi_p} \text{ for } t \geq \xi_p. \quad (2.31)
\end{aligned}
$$

To establish the inequalities (2.31), we note that

$$R(t) = \exp[-H(t)] = \exp\left[-H(\xi_p)\left(\frac{H(t)}{H(\xi_p)}\right)\right]. \quad (2.32)$$

But since $h(t)$ is increasing, it is easily shown (see Section 2.2, Problem 5), that

$$\frac{H(t)}{H(\xi_p)} \leq \frac{t}{\xi_p} \quad \text{for } t \leq \xi_p.$$

Hence, replacing $H(t)/H(\xi_p)$ by t/ξ_p in Equation (2.32) gives

$$R(t) \geq e^{-H(\xi_p)(t/\xi_p)} = (1-p)^{t/\xi_p}. \quad (2.33)$$

Similarly, for $t \geq \xi_p$, one has

$$R(t) \leq (1-p)^{t/\xi_p}. \quad (2.34)$$

This establishes the inequalities (2.31).

It is clear from the proof that the inequalities (2.31) are valid not only for the class of IFR distributions, but also for the less restricted class of life distributions for which $H(t)/t$ is increasing as a function of t. Life distributions for which $H(t)/t$ is increasing are said to have the *increasing hazard rate average* (IHRA) property.

Another result of interest is the following. If the life-time T is IFR with $E(T) = \mu$, then

$$R(t) = P(T > t) \geq e^{-t/\mu}, \quad \text{for all } t < \mu. \quad (2.35)$$

To prove inequality (2.35), we note that the assumption that $h(t)$ is increasing implies that $H(t) = \int_0^t h(x)dx$ is a convex function of t. A function $\varphi(t)$

is said to be convex if $\varphi\left([t_1 + t_2]/2\right) \leq [\varphi(t_1) + \varphi(t_2)]/2$, for any pair of numbers t_1, t_2. Hence, it follows from Jensen's inequality that

$$H[E(T)] \leq E[H(T)] \tag{2.36}$$

and so

$$H(\mu) \leq 1, \tag{2.37}$$

since $U = H(T)$ is a unit-exponential random variable. But,

$$R(t) = P(T > t) = \exp[-H(t)] = \exp\left[-H(\mu)\left(\frac{H(t)}{H(\mu)}\right)\right]. \tag{2.38}$$

Using inequality (2.37) and the inequality

$$\frac{H(t)}{H(\mu)} \leq \frac{t}{\mu} \quad \text{for } t < \mu$$

establishes inequality (2.35).

If $F(t)$ is DFR, then it can be shown analogously that

$$R(t) = P(T > t) \quad \leq \quad (1-p)^{t/\xi_p} \text{ for } t \leq \xi_p$$
$$\geq \quad (1-p)^{t/\xi_p} \text{ for } t \geq \xi_p \tag{2.39}$$

and

$$R(t) = P(T > t) \leq e^{-t/\mu} \text{ for } t < \mu.$$

The inequalities (2.39) are also valid for the less restricted class of life distributions for which $H(t)/t$ is decreasing as a function of t. Life distributions for which $H(t)/t$ is decreasing are said to have the *decreasing hazard rate average* (DHRA) property.

2.3 General remarks on large systems

We wish to mention briefly some results on the failure distributions associated with systems consisting of many components. The most striking feature is that, roughly speaking, the situation becomes simpler as system complexity increases and that, under fairly reasonable conditions, times between system failures are approximately exponentially distributed.

It is important to distinguish two classes of system failures: (1) where repair or replacement of components is possible (i.e., tracking or communications equipment on the ground; computers). In this case, one is particularly interested in the distribution of time intervals between system failures; (2) where repair or replacement is usually not possible (i.e., missiles, unmanned satellites). In this case, the time to first system failure is more important.

Turning first to the repairable case, let us make the simplifying assumption

that any component failure can also be considered a system failure* and that
the time spent in detecting, isolating, and repairing the failure is negligible.†
Let the N components, which comprise the system, be indexed as components, $1, 2, \ldots, N$. Consider component number j and consider the random
sequence of times $T_{j1}, T_{j2}, \ldots$, generated by the original item and its replacements (assuming that failure is detected instantaneously and that a replacement
for the failed item is available instantly). This random sequence of times generates what is known as a "renewal process." Thus we have, so to speak, N
statistically independent renewal processes (one for each component) going on
simultaneously, and the system failure times are generated by the superposition
of these N processes. The properties of pooled renewal processes have been
investigated by Cox and Smith [11] and are described in Chapter 6 of Cox [9]
for the special case where the N statistically independent renewal processes
all have the same failure distribution. If μ is the mean of this p.d.f., Cox and
Smith prove, for large N, that after a long time has elapsed, then the p.d.f.
of the times between successive system failures generated by the N pooled
renewal processes is approximately exponential with mean μ/N. Drenick [16]
generalizes this equilibrium result for the case of large N, to the case where the
N renewal processes associated with the N components and their replacements
are still statistically independent, but may have different component mean failure times μ_i, $i = 1, 2, \ldots, N$. Under conditions given in Drenick's paper
(p. 683), it is shown that, for large N, the equilibrium distribution for times
between successive system failures is approximately exponential with mean
$= (\Sigma_{i=1}^{N} \mu_i^{-1})^{-1}$.

Up to this point, we have considered the asymptotic distribution of times between system failures for a repairable complex system. We pointed out, however, at the start of this discussion, that for some complex systems, one is interested in the distribution of the time to first system failure. Let us again think
of the system as being composed of a large number N of components, and let
us again assume for simplicity that the components operate independently and
that the failure of any component results in a system failure. Let T_j be the time
to failure of the jth component; then clearly system failure occurs at $\min\{T_j\}$,
$j = 1, 2, \ldots, N$. If the components all fail in accordance with the same failure law, the asymptotic distribution of the time to system failure is one of the
extreme value distributions. For a detailed discussion of this theory, see the
classical book of Gumbel [31] or any of the modern books, e.g. Beirlant et al.

---

* In practice, component failures or even subsystem failures need not necessarily result in system
failures. Indeed, the task of the design or systems engineer is frequently to design a system
so that it will continue to operate (possibly with reduced effectiveness) even if components or
subsystems fail. Stochastic models incorporating these more realistic assumptions will be given
elsewhere in this book.

† At the moment, we find it convenient to neglect the time required for repair. The system effectiveness models given later in the book involve both failure and repair times.

[4]. For a review of parts of this theory and its applications, which are particularly relevant in the present context, see Epstein [21]. The key point is that the distribution of the time to system failure will be determined in large part by the behavior of the component failure distribution near $t = 0$. If the common component failure c.d.f. behaves like βt^{α}, for some α, $\beta > 0$ as $t \rightarrow 0^{+}$, then one obtains the Weibull distribution as the system failure law. If $\alpha = 1$, the system failure law is exponential.

Drenick gives a result for the asymptotic distribution of the time to first failure for the more realistic case, where the components do not all fail according to the same failure distribution. Let $F_j(t)$ and $h_j(t)$ be, respectively, the c.d.f. of the times to failure of the jth component and the hazard function of the jth component. Let T_S be the time to first system failure. From the assumption of independence of the N components, it follows that

$$
\begin{aligned}
P(T_S > t) &= P(T_1 > t, \, T_2 > t, \ldots, T_N > t) \\
&= \prod_{j=1}^{N} [1 - F_j(t)] = e^{-\sum_{j=1}^{N} H_j(t)}.
\end{aligned}
\tag{2.40}
$$

As $N \rightarrow \infty$, the behavior of the hazard functions $h_j(t)$ in the neighborhood of $t = 0$ becomes all important in determining the distribution of T_S. This is made precise by Drenick [16], who proves that if

(i) $h_j(t) = h_j(0) + \alpha_j t^{\gamma} + o(t^{\gamma})$ as $t \rightarrow 0$ (with $h_j(0) > 0$, $\gamma > 0$; $j = 1, 2, \ldots, N$).

(ii) $h_S(0) = \sum_{j=1}^{N} h_j(0) \rightarrow \infty$, as $N \rightarrow \infty$.

(iii) $\left| \sum_{j=1}^{N} \alpha_j / h_S(0) \right| < A < \infty$ for all N,

then the random variable $h_S(0)T_S$ converges in distribution to an exponential random variable with rate 1 (or equivalently T_S is, approximately, exponentially distributed with rate $\sum_{j=1}^{N} h_j(0)$).

The conditions of Drenick's theorem are, of course, trivially satisfied if each of the components has an exponential life distribution with mean life θ_j for the jth component. In this case, $h_j(0) \equiv h_j(t) \equiv 1/\theta_j$ and T_S is exponentially distributed with mean life $1/\sum_{j=1}^{N} 1/\theta_j$ (stated in terms of failure rates, this is the familiar statement that the system failure-rate $\Lambda_S = \sum_{j=1}^{N} 1/\theta_j = \sum_{j=1}^{N} \lambda_j$). Of course, the relationship $\lambda_j = 1/\theta_j$ is true and meaningful only for an exponential life distribution.

If the $F_j(t)$ are IFR with corresponding means μ_j, $j = 1, 2, \ldots, N$, it follows from the inequality (2.35) that

$$R_S(t) = P(T_S > t) = \prod_{j=1}^{N}[1 - F_j(t)]$$

$$= \prod_{j=1}^{N} R_j(t) \geq \exp\left[-t\sum_{j=1}^{N}\frac{1}{\mu_j}\right], \quad (2.41)$$

for $t < \min\{\mu_j\}$, $j = 1, 2, \ldots, N$.

This inequality explains why the usual procedure of adding component failure rates frequently results in too conservative a prediction of system reliability. A much better prediction of the reliability of a system composed of a large number N of independent IFR components, meeting the conditions of Drenick's theorem is given by $\exp\left(-\Sigma_{j=1}^{N}h_j(0)t\right)$.

2.4 Problems and comments

Problems for Section 2.1

1. Consider k independent homogeneous Poisson processes each starting at time $t = 0$, having occurrence rates $\{\lambda_i : i = 1, 2, \ldots, k\}$, and operating simultaneously. Let $X_i(t)$ be the number of occurrences generated by the ith process in the time interval $(0, t]$ and let the pooled number of occurrences in $(0, t]$ arising from the superposition of the k Poisson processes by $Y_k(t) = \sum_{i=1}^{k} X_i(t)$.

   (a) Prove that $Y_k(t)$ is Poisson distributed with rate $\Lambda = \sum_{i=1}^{k} \lambda_i$.

   (b) Prove that for $0 \leq n_i \leq n$,

   $$P[X_i(t) = n_i \mid Y_k(t) = n] = \binom{n}{n_i}\left(\frac{\lambda_i}{\Lambda}\right)^{n_i}\left(1 - \frac{\lambda_i}{\Lambda}\right)^{n-n_i}.$$

   (c) Prove that the random vector $(X_1(t), X_2(t), \ldots, X_k(t))$ has a multinomial distribution, namely

   $$P\left[X_1(t) = n_1, X_2(t) = n_2, \ldots, X_k(t) = n_k \mid Y_k(t) = \sum_{i=1}^{k} n_i = n\right]$$

   $$= \frac{n! \prod_{i=1}^{k}(\lambda_i/\Lambda)^{n_i}}{\prod_{i=1}^{n}(n_i!)}.$$

   (d) Let $T_{1,k}$ be the time to the first occurrence in the pooled process. Prove that $T_{1,k}$ is exponentially distributed with p.d.f. $\Lambda e^{-\Lambda t}$, $t \geq 0$. What is the p.d.f. of the time between the jth and $(j + 1)^{\text{th}}$ occurrence in the pooled process?

2. The failure-rate of a certain type of transistor is assumed to be a constant $\lambda = 10^{-6}$/hour. A certain system contains 2,500 of these transistors. Failed transistors are immediately replaced by good transistors, so as to keep the total number of transistors = 2,500.

   (a) What is the expected number of transistor failures in 10,000 hours of usage?

   (b) What is the probability that the number of transistors failing in 10,000 hours exceeds 30? Is less than 15?

   Hint: Use the results of Problem 1 and the central limit theorem, since for Λt "large" (say greater than 25), $Y_k(t)$ is approximately normally distributed with mean Λt and variance Λt.

3. Consider a situation where an item is exposed simultaneously to two independent competing causes for failure. If the item fails because of the first (second) cause, it is said to be a Type 1(2) failure. Let us assume a simple peak model for each type of failure with peaks generated by Poisson processes having rates λ_1 and λ_2, respectively. Assume that a failed item is replaced immediately by a new one.

   (a) Prove that the (conditional) probability of a Type 1 failure is given by $\lambda_1/(\lambda_1 + \lambda_2)$.

   (b) Prove that the p.d.f. of failure-times, where no distinction is made as to type of failure, is $(\lambda_1 + \lambda_2)e^{-(\lambda_1+\lambda_2)t}$, $t \geq 0$.

   (c) Prove that the conditional p.d.f. of the failure-times, given all failures are of Type 1 (or of Type 2), is also

   $$(\lambda_1 + \lambda_2)e^{-(\lambda_1+\lambda_2)t}, \qquad t \geq 0.$$

4. Generalize Problem 3 to the case of k competing causes for failure having rates λ_i, $i = 1, 2, \ldots, k$.

5. Consider two machines, where failures from machine 1(2) occur with constant failure rates $\lambda_1(\lambda_2)$. During the first 5,000 hours of operation of each machine, a total of 20 failures was observed with 15 failures arising from machine 1 and 5 failures from machine 2. What can you say about the hypothesis that both machines have the same failure rate?

   Hint: Under the hypothesis that $\lambda_1 = \lambda_2$, X_1, the number of failures arising from machine 1, given that there were a total of 20 failures from both machines, is binomially distributed with $n = 20$ and $p = 1/2$. Testing the hypothesis $\lambda_1 = \lambda_2$ against the alternative $\lambda_1 \neq \lambda_2$ is equivalent to testing $p = 1/2$ against $p \neq 1/2$. The hypothesis $p = 1/2$ becomes less and less tenable the more X_1 differs from 10, the expected number of failures if $\lambda_1 = \lambda_2$. In this problem, the observed number of failures = 15, which differs from the expected number by 5. This is fairly strong evidence against the hypothesis that $\lambda_1 = \lambda_2$ since $P(|X_1 - 10| \geq 5) = 43400/2^{20} = .04139 < .05$.

6. Suppose that 100 failed items are generated in accordance with the model described in Problem 3. It is found that 40 of the failures are Type 1 and 60 are Type 2. On the basis of the above evidence, would you be willing to accept the assumption that $\lambda_2 = 4\lambda_1$? Sketch the relevant statistical arguments.

7. Suppose that we have n devices. One by one these devices are exposed to two independent competing causes for failure occurring with rates λ_1 and λ_2, respectively. The failure-time of the jth device is t_j and the Type 1 indicator is δ_j, $j = 1, 2, \ldots, n$. Let $n_1 = \Sigma \delta_j$ be the number of Type 1 failures and $n_2 = n - n_1$. Let $T_n = \sum_{j=1}^{n} t_j$.

   (a) Show that the pair (T_n, n_1) is sufficient for the pair (λ_1, λ_2).

   (b) Find the MLE and UMVUE of (λ_1, λ_2).

8. Show that the p.d.f. $g(t)$ in Equation (2.6) can be written as

$$g(t) = \sum_{k=1}^{\infty} \alpha_k f_k(t),$$

where $\alpha_k = q^{k-1}p$ and

$$f_k(t) = \frac{\lambda(\lambda t)^{k-1} e^{-\lambda t}}{(k-1)!}, \qquad t \geq 0, \quad k = 1, 2, \ldots.$$

9. Consider the shock model in which shocks occur with rate λ and where the conditional probability that the device fails (does not fail) given that a shock occurs is $p(q = 1 - p)$. Assume that the device is instantaneously repaired and returned to service, if it fails. Suppose that in the interval of observation $(0, t]$, $N_1(t)$ failures and $N_2(t)$ nonfailures occur. Prove that $\{N_1(t)\}$ and $\{N_2(t)\}$ are independent homogeneous Poisson processes with rates λp and λq, respectively.

10. Suppose that a device is subject to shocks, where the times between shocks are independent random variables with common p.d.f. $f_2(t) = \lambda^2 t e^{-\lambda t}$, $t \geq 0$. Suppose that the conditional probability that the device fails (does not fail) given that a shock occurs is $p(q = 1 - p)$. It is assumed that the device has no memory, i.e., if it has survived $k - 1$ shocks, then the conditional probability of surviving the kth shock remains equal to p. Let T be the random variable for the length of time until the device fails.

    (a) Show that $g(t)$, the p.d.f. of T, is given by

$$g(t) = \frac{\lambda p}{2\sqrt{q}} \left\{ e^{-\lambda(1-\sqrt{q})t} - e^{-\lambda(1+\sqrt{q})t} \right\}, \qquad t \geq 0.$$

    (b) Suppose that $\lambda = .1/\text{hour}$ (i.e., the mean of $f_2(t)$ is 20 hours) and $p = .1$. Find the probability that the device survives 100 hours, 200 hours, 400 hours, 600 hours.

(c) Show that as $p \to 0$, the c.d.f. of the random variable $\lambda p T/2$ approaches $1 - e^{-t}$, $t \geq 0$.

Comment. Stochastic models involving exposure to several competing causes for failure have been of interest for a long time to actuaries and biostatisticians. After all, human beings are at all times subject simultaneously to many essentially independent risks of injury or death. An important problem in this context is the estimation of the mortality rates due to a particular cause, when not only this cause but many others are operating simultaneously. The interested reader is referred to books such as David and Moeschberger [14], Crowder [12] and Pintilie [42].

11. Construct examples where the two models in Equations (2.10) and (2.12) might be a reasonable idealization of the true situation.

12. Generalize the alternative model by replacing p by $p(t)$.

13. Consider a competing risks model in which there are two independent causes of failure for a device. Failures due to cause i are generated by a time-dependent Poisson process having rate $\lambda_i(t)$, $i = 1, 2$. Suppose that $\lambda_2(t) = c\lambda_1(t)$.

   (a) Prove that the conditional probability of a Type 1 failure is given by $1/(1 + c)$.

   (b) Prove that the p.d.f. of failure times of all devices, as well as of those having a specific failure type, is

   $$(1 + c)\, \lambda_1(t)\, e^{-(1+c)\Lambda_1(t)}, \quad t \geq 0,$$

   where $\Lambda_1(t) = \int_0^t \lambda_1(\tau)d\tau$.

14. Consider the shock model in which shocks occur with rate λ and where the conditional probability that the device fails (does not fail) given that a shock occurs is $p(t)(q(t) = 1 - p(t))$. Assume that the restoration time, if failure occurs, is negligible. Suppose that in the interval of observation $(0, t]$, $N_1(t)$ failures and $N_2(t)$ nonfailures occur. Show that $\{N_1(t)\}$ and $\{N_2(t)\}$ are independent Poisson processes with mean measures $\lambda P(t)$ and $\lambda[t - P(t)]$, respectively.

15. Consider the shock model in which shocks occur with rate $\lambda(t)$ and where the conditional probability that the device fails (does not fail) given that a shock occurs is $p(q = 1 - p)$. Assume that the restoration time, if failure occurs, is negligible. Suppose that in the interval of observation $(0, t]$, $N_1(t)$ failures and $N_2(t)$ nonfailures occur. Show that $\{N_1(t)\}$ and $\{N_2(t)\}$ are independent Poisson processes with mean measures $p\Lambda(t)$ and $q\Lambda(t)$, respectively.

16. An item has a coating of thickness u. From time to time, it is struck by a

blow. It is assumed that the blows occur according to a homogeneous Poisson process with rate λ. Suppose that each blow chips off material having exactly thickness Δ.

(a) What is the probability that the coating is chipped off by time t?

(b) Compute the numerical value of this probability if $u = .01$ inch, $\Delta = .001$ inch, $\lambda = .01/\text{hour}$ and $t = 1,500$ hours.

(c) Let T be the time when the coating is chipped off. For the values u, Δ, and λ in (b), find the p-quantile t_p of T for $p = .05, .10, .50, .90, .95$.

17. Suppose that the coating on the tungsten filament of a bulb is .01 inch thick. Suppose that each time the bulb is turned on the thickness of the coating chipped off is a random variable X with p.d.f. $\frac{1}{\Delta}e^{-x/\Delta}$, $x \geq 0$, with $\Delta = .0001$ inch.

What is the probability that the bulb can be turned on 80 times without failure; 100 times without failure; 120 times without failure? Graph the probability of bulb failure (on or before n) as a function of n, the number of times that the bulb is turned on.

18. Assume in Problem 17 that the successive times when the bulb is turned on can be considered as being generated by a Poisson process with rate $\lambda = .1/\text{hour}$. What is the probability that the bulb will last 800 hours; 1,000 hours; 1,200 hours? Graph the probability that the bulb will fail (on or before time t) as a function of t.

Hint: Let Y_t be the thickness of the coating chipped off by time t. For t "large" it can be shown that Y_t is approximately normally distributed with mean $\mu_t = \lambda\Delta t$ and variance $\sigma_t^2 = 2\lambda\Delta^2 t$. See, e.g., Cox [9].

Problems for Section 2.2

1. Suppose that the underlying p.d.f. of life is a mixture of two exponential distributions, given by

$$f(t) = \alpha\lambda_1 e^{-\lambda_1 t} + (1-\alpha)\lambda_2 e^{-\lambda_2 t}, \qquad t \geq 0,$$

with

$$0 < \alpha < 1, \quad \lambda_2 \neq \lambda_1.$$

Prove that the hazard function $h(t)$ is a decreasing function of t.

Comment. More generally, it can be shown that any convex combination of exponential p.d.f.s of the form $f(t) = \sum_{i=1}^{\infty} \alpha_i \lambda_i e^{-\lambda_i t}$, $\alpha_i \geq 0$, $\sum_{i=1}^{\infty} \alpha_i = 1$ is DFR. This in turn is a special case of the result proved in Barlow and Proschan [3] that if $F_i(t)$ is a DFR distribution for each $i = 1, 2, \ldots, \alpha_i \geq 0$ for each $i = 1, 2, \ldots$, and $\sum_{i=1}^{\infty} \alpha_i = 1$, then $G(t) = \sum_{i=1}^{\infty} \alpha_i F_i(t)$ is a DFR distribution.

2. Suppose that the life T of a system is $\Gamma(a, \lambda)$. Prove that the hazard function $h(t)$ is an increasing function of t, if $a > 1$, and is a decreasing function of t, if $0 < a < 1$.

3. Suppose that the underlying p.d.f. of life is truncated normal distribution

$$
\begin{aligned}
f(t) &= \frac{1}{\sqrt{2\pi}\sigma A} e^{-(t-t_0)^2/2\sigma^2} \quad , \quad \text{if } t \geq 0 \\
&= 0 \quad\quad\quad\quad\quad\quad\quad\quad\quad , \quad \text{if } t < 0,
\end{aligned}
$$

where $t_0 \geq 0$ and

$$
A = \int_0^\infty \frac{1}{\sqrt{2\pi}\sigma} e^{-(t-t_0)^2/2\sigma^2} \, dt .
$$

Show that $h(t)$ is a monotonically increasing function of t.

4. Prove the assertion that $U = H(T)$ is distributed with p.d.f. e^{-u}, $u \geq 0$. How is $2U$ distributed?

5. Show that $h(t)$ increasing (decreasing) implies that $H(t)/t$ is increasing (decreasing) (this means that IFR implies IHRA and DFR implies DHRA).

6. Suppose that $F(t)$ is IFR (DFR). Show, using the result of Problem 5 that $(R(t))^{1/t}$ is decreasing (increasing). Note that this means that $(R(t))^{1/t}$ is decreasing (increasing) in t if $F(t)$ is IHRA (DHRA).

7. Let T be the life time of a device with hazard function $h(t)$. Prove that for $\tau > 0$,

$$
P(T > t + \tau \mid T > t) = e^{-[H(t+\tau)-H(t)]}.
$$

In other words: Suppose that a device has survived for a length of time t. What is the (conditional) probability that the device will survive an additional τ time-units?

8. (Continuation of Problem 7). Suppose that $h(t)$ is an increasing (decreasing) function of t. Show that $P(T > t + \tau \mid T > t)$ is for fixed $\tau > 0$ decreasing (increasing) in t. Conversely, show that if $P(T > t + \tau \mid T > t)$ is for fixed $\tau > 0$ decreasing (increasing) in t, then $h(t)$ is increasing (decreasing).

9. Consider a continuous IHRA distribution $F(t)$ with known p-quantile ξ_p (for a fixed $0 < p < 1$).

   (a) Verify that the exponential distribution with mean $\theta_p = -\xi_p/\log(1-p)$ has the same p-quantile.

   (b) Suppose that we draw $R(t) = 1 - F(t)$ and $R^*(t) = e^{-t/\theta_p}$ on the same graph with t as abscissa. What is the graphical interpretation of the inequality (2.31)? Which has the better chance of surviving for a length of time $t < \xi_p$, an item with life time c.d.f. $F(t)$ or an item whose life time is exponentially distributed with mean θ_p?

10. Prove the implications

$$
IFR \Rightarrow IFRA \Rightarrow NBU
$$

and

$$DFR \Rightarrow DFRA \Rightarrow NWU.$$

11. Consider an IFR distribution with c.d.f. $F(t)$ having mean θ and an exponential distribution $G(t) = 1-e^{-t/\theta}$, $t \geq 0$. Suppose that $R_1(t) = 1-F(t)$ and $R_2(t) = 1 - G(t) = e^{-t/\theta}$ are both plotted on the same graph with t as the common abscissa.

   (a) Show that $R_1(t)$ either coincides with $R_2(t)$ or crosses it exactly once from above.

   (b) What does the inequality (2.35) tell us about the location of the intersection of $R_1(t)$ and $R_2(t)$?

12. Consider an IHRA (DHRA) life distribution with c.d.f. $F(t) = 1 - e^{-H(t)}$, $t \geq 0$.

   (a) Show that $H(\alpha t) \leq (\geq) \alpha H(t)$ and $R(\alpha t) \geq (\leq)[R(t)]^{\alpha}$ for $0 \leq \alpha \leq 1$;

   (b) Show that $H(\beta t) \geq (\leq) \beta H(t)$ and $R(\beta t) \leq (\geq) [R(t)]^{\beta}$ for $\beta \geq 1$.

13. A device whose life time follows an IHRA distribution is known to survive 100 hours with probability .9. What can be said about the probability of the device surviving 10 hours? 50 hours? 200 hours? What can be said about the conditional probability that a device survives 300 hours given that it survived 100 hours? Answer the same questions if the word IHRA is replaced by DHRA.

14. Suppose that one starts with a population of N_0 persons of zero age at $t = 0$. Let N_t be the number of individuals surviving after t years. Give an approximate formula for the "force of mortality" $h(t)$, during the one-year time interval $(t, t + 1]$.

15. Suppose n items, each having hazard function $h(t)$, are placed on test at time $t = 0$. Items that fail are not replaced. Consider the stochastic process generated by successive failures and the associated random function $N(t)$, the number of failures in $(0, t]$. Let $P_k(t) = P[N(t) = k]$.

   (a) Verify that

   $$P_k(t) = \binom{n}{k}[1 - e^{-H(t)}]^k e^{-(n-k)H(t)}, \quad k = 0, 1, 2, \ldots, n.$$

   (b) Show that $T_{k,n}$, the time to the kth failure, is distributed with p.d.f.

   $$\frac{n!}{(k-1)!(n-k)!} h(t)[1 - e^{-H(t)}]^{k-1} e^{-(n-k+1)H(t)}, \quad t \geq 0.$$

16. Suppose n items, each having hazard function $h(t)$, are placed on test at time $t = 0$. Any item that fails is replaced by an item of the same "age" as the item that has failed.

(a) Show that the successive times to failure are generated by a nonhomogeneous Poisson process with rate $\lambda(t) = nh(t)$.

(b) Using the results of (a), find the probability distribution of $N(t)$ the number of failures in $(0, t]$ and also the p.d.f. of T_k, the time to the kth failure.

(c) Show that the expected number of failures in $(0, t]$ is given by $E[N(t)] = nH(t)$.

17. Suppose that a device can fail in two competing and independent ways (called failures of type 1 and 2, respectively). Let the time to a failure of type i be described by a random variable T_i, with associated c.d.f., p.d.f., and hazard functions $F_i(t)$, $f_i(t)$, $h_i(t)$, respectively, $i = 1, 2$. The random variables T_1 and T_2 are assumed to be independent.

(a) Show that the p.d.f. of the time to failure, T, of the device is given by $[h_1(t) + h_2(t)] e^{-[H_1(t) + H_2(t)]}$, $t \geq 0$, where $H_i(t) = \int_0^t h_i(\tau) d\tau$, $i = 1, 2$.

(b) Prove that the conditional probability that the failure is of type i given that the device fails at time t is given by

$$\frac{h_i(t)}{h_1(t) + h_2(t)}, \qquad i = 1, 2.$$

(c) Prove that this ratio is a constant independent of t if and only if $1 - F_1(t) = [1 - F_2(t)]^c$, where c is a positive constant.

(d) Prove that if $F_1(t)$ and $F_2(t)$ are related by the relation in (c), then the probability that a failure will be of type 1 is given by $c/(c+1)$.

18. (Continuation of Problem 17). Prove that the probability that the failure is of type 1(2) and occurs at or before time t, is given by

$$G_1(t) = \int_0^t f_1(x)[1 - F_2(x)]dx = \int_0^t h_1(x) \prod_{i=1}^{2} [1 - F_i(x)]dx$$

and

$$G_2(t) = \int_0^t f_2(x)[1 - F_1(x)]dx = \int_0^t h_2(x) \prod_{i=1}^{2} [1 - F_i(x)]dx.$$

Also, prove that

$$F_i(t) = 1 - \exp\left\{ -\int_0^t \left(1 - \Sigma_{j=1}^2 G_j(x)\right)^{-1} g_i(x)dx \right\},$$

where $g_i = G_i'$.

19. Suppose that life-times T of certain items are distributed with continuous p.d.f. $f(t)$.

(a) Show that the (conditional) p.d.f. of X_τ, the life-time of those items which have survived for a length of time τ is given by $f(x)/(1 - F(\tau))$, $x \geq \tau$.

(b) Define $X_\tau - \tau$ as the residual life of those items which have survived for a length of time τ and $E(X_\tau - \tau)$ as the mean residual life. Evaluate the mean residual life for the case where $f(t) = \lambda^3 t^2 e^{-\lambda t}/2$, $t \geq 0$, and show that it is decreasing as a function of τ.

20. Suppose n "new" items, each having hazard function $h(t)$, are placed on test at time $t = 0$. The test is continued until time t^* with any failed items being replaced by an item of the same "age" as the one that failed. Let $N(t^*)$ be the number of items that have failed by time t^*. Show that $(1 - 1/n)^{N(t^*)}$ provides an unbiased estimate of the probability that an item will survive time t^*.

Hint: Use the Rao-Blackwell method.

21. A strand of yarn consists of N_0 fibers, each of which, independent of the other fibers, is subject to failure under load. Failure occurs when all the fibers have failed. Suppose that the probability of a single fiber failing in the time interval $(t, t + \Delta t]$ is given by $h(t)\Delta t + o(\Delta t)$ independently of how many fibers have failed by time t. Let $T = $ life time of the strand. Show that

(a) $F(t) = P(T \leq t) = \left[1 - e^{-H(t)}\right]^{N_0}$ where $H(t) = \int_0^t h(x)dx$.

(b) If $h(t) \equiv \lambda$ for all $t > 0$, show that the p.d.f. of T is given by

$$f(t) = N_0 \lambda e^{-\lambda t} \left(1 - e^{-\lambda t}\right)^{N_0 - 1}, \qquad t \geq 0.$$

Also show that

$$E(T) = \frac{1}{\lambda}\left[1 + \frac{1}{2} + \cdots + \frac{1}{N_0}\right].$$

22. (Continuation of Problem 21). Suppose that the probability of a single fiber failing in $(t, t + \Delta t]$ when n fibers are left is given by $(N_0/n)h(t)\Delta t + o(\Delta t)$. Let $T = $ life time of the strand. Show that

(a)

$$F(t) = P(T \leq t) = 1 - \sum_{j=0}^{N_0 - 1} [N_0 H(t)]^j e^{-N_0 H(t)}/j!.$$

(b) If $h(t) \equiv \lambda$ for all $t > 0$, show that the p.d.f. of T is given by

$$f(t) = \frac{N_0 \lambda (N_0 \lambda t)^{N_0 - 1} e^{-N_0 \lambda t}}{(N_0 - 1)!}, \qquad t \geq 0.$$

Also show that $E(T) = 1/\lambda$.

Note, in this model, the hazard rate of the system is $N_0 h(t)$ as long as the system is still alive.

Problems for Section 2.3

1. Consider a life time distribution $F(t)$, possessing finite mean μ, and monotonically increasing hazard function $h(t)$. Prove that $h(0) \leq \mu^{-1}$.

2. Use the result of Problem 1 to prove the assertion that for a system composed of N independent IFR components,

$$\exp\left(-t\Sigma_{j=1}^{N}h_j(0)\right) \geq P(T_S > t) \geq \exp\left(-t\Sigma_{j=1}^{N}\mu_j^{-1}\right) .$$

CHAPTER 3

Reliability of various arrangements of units

3.1 Series and parallel arrangements

3.1.1 Series systems

Let us consider a system S composed of n subsystems $S_1, S_2, \ldots, S_n$ which are all arranged in a series (see Figure 3.1). It is assumed that the $\{S_i\}$ fail independently and must all operate successfully if the system S is to operate successfully.

Figure 3.1 *A series system with n components.*

Let $F_i(t)$ be the c.d.f. of the time to failure of S_i. We wish to find $G_s(t)$, the c.d.f. of system failure time. Let T_i be the time to failure of the ith subsystem and let $T = $ time to failure of the system. Then, clearly the event $T > t$ (the system survives for a length of time t) occurs if and only if all $T_i > t$ (all subsystems survive time t), i.e.,

$$T = \min_i \{T_i\}$$

and

$$P(T > t) = P(T_1 > t,\ T_2 > t, \ldots, T_n > t). \qquad (3.1)$$

But, by hypothesis, the events $T_i > t$ are mutually independent. Hence,

$$P(T > t) = \prod_{i=1}^{n} P(T_i > t) = \prod_{i=1}^{n}(1 - F_i(t)). \qquad (3.2)$$

Hence, $G_s(t)$, the c.d.f. of system failure time, becomes

$$G_s(t) = 1 - P(T > t) = 1 - \prod_{i=1}^{n}(1 - F_i(t)). \tag{3.3}$$

The reliability of the system for the time interval $(0, t]$ is defined as the probability that the system survives for a length of time t and is denoted by $R_s(t) = P(T > t)$. Similarly, the reliability for each subsystem is given by $R_i(t) = P(T_i > t)$. Hence (3.2) can be rewritten as

$$R_s(t) = \prod_{i=1}^{n} R_i(t). \tag{3.4}$$

In particular, if the n subsystems all have a common time to failure c.d.f. $F(t)$, then $G_s(t)$ becomes

$$G_s(t) = 1 - [1 - F(t)]^n \tag{3.5}$$

and

$$R_s(t) = [1 - F(t)]^n. \tag{3.6}$$

Example. Suppose that the life distribution of S_i is exponential with mean $\theta_i = 1/\lambda_i$. Then $F_i(t) = 1 - e^{-\lambda_i t}$ and from Equation (3.3)

$$G_s(t) = 1 - \prod_{i=1}^{n}(1 - F_i(t)) = 1 - e^{-\sum_{i=1}^{n} \lambda_i t}.$$

Hence, the system life is also exponential with mean

$$\Theta_s = \frac{1}{\sum_{i=1}^{n} \lambda_i} = \frac{1}{\sum_{i=1}^{n} \theta_i^{-1}}.$$

The system failure rate

$$\Lambda_s = \frac{1}{\Theta_s} = \sum_{i=1}^{n} \lambda_i.$$

The system reliability $R_s(t) = e^{-\Lambda_s t}$.

3.1.2 Parallel systems

Let us consider a system S composed of n subsystems $S_1, S_2, \ldots, S_n$ are arranged in parallel redundancy, i.e., are all operating simultaneously, as in Figure 3.2.

It is assumed that the S_j fail independently of each other and that the system S operates successfully if at least one of the S_j operates successfully.

Let $F_j(t)$ be the c.d.f. of the time to failure of S_j. We wish to find $G_p(t)$, the

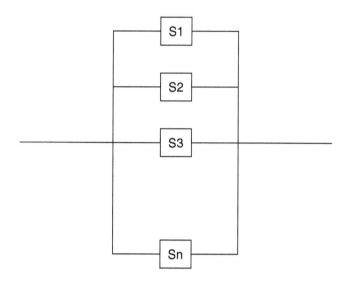

Figure 3.2 *A parallel system with n components.*

c.d.f. of system failure-time. Let T_j be the time to failure of S_j and let T be the time to failure of the system. Then, clearly

$$T = \max_j \{T_j\},$$

thus, due to the independence,

$$G_p(t) = P\{T \le t\} = \prod_{j=1}^{n} P\{T_j \le t\} = \prod_{j=1}^{n} F_j(t).$$

The reliability of the system over the time interval $(0, t]$ becomes

$$R_p(t) = 1 - G_p(t) = 1 - \prod_{j=1}^{n} F_j(t). \tag{3.7}$$

In particular, if the n subsystems all have a common time to failure c.d.f. $F(t)$, then

$$G_p(t) = F^n(t) \tag{3.8}$$

and

$$R_p(t) = 1 - F^n(t). \tag{3.9}$$

Example. Consider a system of n units in parallel redundancy, with each unit having a common time to failure exponential c.d.f. $F(t) = 1 - e^{-\lambda t}$, $t \ge 0$, and system reliability $R_p(t) = 1 - (1 - e^{-\lambda t})^n$. Unlike the series system, here the system life is not exponential (for $n \geq 2$). The system hazard rate $h_p(t)$ is

$$h_p(t) = \frac{G'_p(t)}{R_p(t)} = \frac{n\lambda e^{-\lambda t}(1 - e^{-\lambda t})^{n-1}}{1 - (1 - e^{-\lambda t})^n}.$$

It is easily verified that h_p is increasing from $h_p(0) = 0$ to $h_p(\infty) = \lambda$, which means, that given the system has survived for a long time t, it will continue to operate, approximately, as a one-unit system with failure rate λ. This is to be contrasted with the constant failure rate of each individual unit.

The mean time to system failure is given by

$$ET = \frac{1}{\lambda}\left(1 + \frac{1}{2} + \cdots + \frac{1}{n}\right). \tag{3.10}$$

This represents a modest increase in the mean time to failure when compared with mean life $1/\lambda$ of a single unit. The principal benefit obtained from redundancy is the reduction in unreliability for sufficiently small λt (say $\lambda t < .1$). Thus, for the case $n = 3$, and $\lambda t = .05$, the system unreliability would be 1.16×10^{-4} as compared with 5×10^{-2} for a single unit by itself. Thus, the unreliability has been reduced by a factor of 420 although the expected time to system failure has been increased by a factor of $11/6$.

One occasionally sees reliability analyses in which system consisting of n units in parallel redundancy, each unit having failure rate λ, is considered as if it were a single unit having constant failure rate $\lambda^* = 1/ET = \lambda/\Sigma j^{-1}$. It is clear from our discussion that this is wrong and will result in large errors. For instance, in the case where $n = 3$ and $\lambda t = .05$, a single unit having constant failure rate $\lambda^* = \lambda 6/11$ would have unreliability $2.69 \times !0^{-2}$ as compared with 1.16×10^{-4}, the unreliability of the redundant system.

3.1.3 The k out of n system

A generalization which includes both the series and parallel arrangements, is the so-called k out of n system. It is assumed that the system is composed of n units, each of which has the same failure time distribution $F(t)$, and where the units are assumed to fail independently of each other. The system is said to operate successfully over the time interval $(0, t]$ if at least k out of the n units operate successfully over $(0, t]$. In accordance with this definition, a series system is an n out of n system (i.e., $k = n$) and a parallel system is a one out of n system (i.e., $k = 1$).

If $T_1 \leq T_2 \leq \cdots \leq T_n$ are the failure-times of the single units, then the T_j are the order statistics of a sample of size n from F. On the other hand, the system life T is exactly T_{n-k+1}. Using the results on order statistics in Subsection

1.3.8, the system reliability $R_{k,n}(t)$ is given by

$$R_{k,n}(t) = P\{T_{n-k+1} > t\} = \sum_{j=k}^{n} \binom{n}{j} [1 - F(t)]^j [F(t)]^{n-j}. \qquad (3.11)$$

In agreement with our previous discussion, $R_{n,n}(t)$, the reliability of a system consisting of n units in series, is given by

$$R_{n,n}(t) = P\{T_1 > t\} = [1 - F(t)]^n \qquad (3.12)$$

and $R_{1,n}$, the reliability of n units in parallel redundancy, is given by

$$R_{1,n}(t) = P\{T_n > t\} = \sum_{j=1}^{n} \binom{n}{j} [1 - F(t)]^j [F(t)]^{n-j} = 1 - F^n(t). \quad (3.13)$$

Frequently we are particularly interested in $R_{k,n}(t)$, evaluated at a specific value $t = t^*$. Denote by $r = 1 - F(t^*)$ the reliability of a single unit over the time interval $(0, t^*]$. Then, Equations (3.11)–(3.13) become, respectively,

$$
\begin{aligned}
R_{k,n}(t^*) &= \sum_{j=k}^{n} \binom{n}{j} r^j (1-r)^{n-j}, \\
R_{n,n}(t^*) &= r^n && \text{(series system)}, && (3.14)\\
R_{1,n}(t^*) &= 1 - (1-r)^n && \text{(parallel system)}.
\end{aligned}
$$

Thus the computation of the reliability for a k out of n system for fixed t^* involves the binomial distribution.

Analogously, if we define $r_i = 1 - F_i(t^*)$, Equation (3.4) becomes

$$R_s(t^*) = \prod_{i=1}^{n} r_i \qquad (3.15)$$

and Equation (3.7) becomes

$$R_p(t^*) = 1 - \prod_{j=1}^{n} (1 - r_j). \qquad (3.16)$$

3.2 Series-parallel and parallel-series systems

By a *series-parallel system* we mean a system composed of n subsystems in series, where subsystem i consists of m_i elements in parallel, $i = 1, 2, \ldots, n$. Assume all elements are active and assume complete independence of all elements. Let A_{ij} be the jth element ($j = 1, 2, \ldots, m_i$) in the ith subsystem and let F_{ij} be the c.d.f. of the time to failure T_{ij} of element A_{ij}. Suppose that the system functions successfully if and only if all subsystems function success-fully. Further, suppose that each subsystem functions successfully if at least one of its elements functions successfully. It follows that the system life T is given by

$$T = \min_i\{T_i\} = \min_i\{\max_j\{T_{ij}\}\},$$

where $T_i = \max_j\{T_{ij}\}$ is the life of subsystem i^{th} subsystem, denoted by A_i.

We wish to find $R_{sp}(t)$, the reliability of the series-parallel system over the time interval $(0, t]$. By Equation (3.7) one has

$$R_{A_i}(t) = P\{T_i > t\} = 1 - \prod_{j=1}^{m_i} F_{ij}(t), \quad i = 1, 2, \ldots, n. \tag{3.17}$$

However, the series-parallel system is composed of the n subsystems $\{A_1, A_2, \ldots, A_n\}$ in series. Hence, by Equations (3.4) and (3.17) one has

$$R_{sp}(t) = P\{T > t\} = \prod_{i=1}^{n} R_{A_i}(t) = \prod_{i=1}^{n}\left(1 - \prod_{j=1}^{m_i} F_{ij}(t)\right). \tag{3.18}$$

If we are interested in the reliability of the series-parallel system for some specific time t^*, let us denote the reliability of A_{ij} over the time interval $(0, t^*]$ as $r_{ij} = 1 - F_{ij}(t^*)$. Then,

$$r_{sp} = R_{sp}(t^*) = \prod_{i=1}^{n}\left(1 - \prod_{j=1}^{m_i}(1 - r_{ij})\right). \tag{3.19}$$

If the elements of A_i have a common reliability $r_{ij} \equiv r_i$, then Equation (3.19) becomes

$$r_{sp} = \prod_{i=1}^{n}\left(1 - (1 - r_i)^{m_i}\right). \tag{3.20}$$

If, in addition, $r_i \equiv r$ and $m_i \equiv m$, then Equation (3.20) becomes

$$r_{sp} = \left(1 - (1 - r)^m\right)^n. \tag{3.21}$$

By a *parallel-series system* we mean a system composed of m subsystems in parallel, where each subsystem A_j consists of n_j elements in series, $j = 1, 2, \ldots, m$. Assume that all elements are active and assume complete inde-pendence of all elements. Let A_{ji}, $i = 1, 2, \ldots, n_j$, be the ith element in A_j and let $F_{ji}(t)$ be the c.d.f. of the time to failure T_{ji} of the element A_{ji}. Sup-pose that the system functions successfully if at least one of the subsystems functions successfully. Suppose further that a given subsystem functions suc-cessfully if and only if all its elements function successfully.

We wish to find $R_{ps}(t)$, the reliability of the parallel-series system for the time interval $(0, t]$. From Equation (3.4) we can say that $R_{A_j}(t)$, the reliability of subsystem A_j is given by

$$R_{A_j}(t) = \prod_{i=1}^{n_j}(1 - F_{ji}(t)), \quad j = 1, 2, \ldots, m. \tag{3.22}$$

But the parallel-series system is composed of the m subsystems $\{A_1, A_2, \ldots, A_m\}$ in parallel. Hence, by Equations (3.7) and (3.22) we have

$$R_{ps}(t) = 1 - \prod_{j=1}^{m}(1 - R_{A_j}(t)) = 1 - \prod_{j=1}^{m}\left(1 - \prod_{i=1}^{n_j}[1 - F_{ji}(t)]\right). \tag{3.23}$$

If we are interested in the reliability of the parallel-series system for some specific time t^*, we define as before $r_{ji} = 1 - F_{ji}(t^*)$ and Equation (3.23) becomes

$$r_{ps} = R_{ps}(t^*) = 1 - \prod_{j=1}^{m}\left(1 - \prod_{i=1}^{n_j}r_{ji}\right). \tag{3.24}$$

If $r_{ji} \equiv r_j$, then Equation (3.24) becomes

$$r_{ps} = 1 - \prod_{j=1}^{m}\left(1 - r_j^{n_j}\right). \tag{3.25}$$

If, in addition, $r_j \equiv r$ and $n_j \equiv n$, then Equation (3.25) becomes

$$r_{ps} = 1 - (1 - r^n)^m. \tag{3.26}$$

Example. Suppose we have to design a system which needs for a successful operation two components in series: a capacitor and a resistor (say A and B, respectively). Suppose further that to increase the reliability of the system it is possible (financially and physically) to use two capacitors and two resistors. We consider two possible arrangements: the series-parallel and the parallel-series, as in Figure 3.3.

Assume that the reliability of each capacitor (transistor) for the specified period of operation is r_A (r_B). Which arrangement is preferred? Well, by Equation (3.21) we have

$$r_{sp} = \left(1 - (1 - r_A)^2\right)\left(1 - (1 - r_B)^2\right)$$

and by Equation (3.26)

$$r_{ps} = 1 - (1 - r_A r_B)^2.$$

It is not hard to show that

$$r_{sp} - r_{ps} = 2r_A(1 - r_A)r_B(1 - r_B), \tag{3.27}$$

which is strictly positive (except for trivial, nonpractical cases). Hence, the

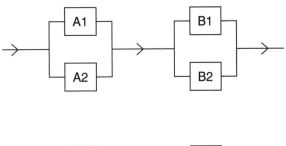

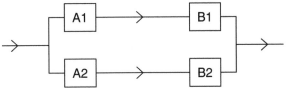

Figure 3.3 *A series-parallel system (upper diagram) and a parallel-series system (lower diagram), both with m = n = 2.*

series-parallel arrangement is preferred. For instance, if $r_A = .95$ and $r_B = .9$ we have

$$r_{sp} = \left(1 - .05^2\right)\left(1 - .1^2\right) = .987525$$

and

$$r_{ps} = 1 - (1 - .95 \times .9)^2 = .978975.$$

Similarly, if $r_A = r_B = .9$, we have

$$r_{sp} = \left(1 - (1 - .9)^2\right)^2 = .9801$$

and

$$r_{ps} = 1 - \left(1 - .9^2\right)^2 = .9639.$$

3.3 Various arrangements of switches

One might get the impression from the previous discussion that putting units in parallel always increases reliability. This need not be the case. Consider a parallel array of mechanical switches (or diodes, if electronic). Suppose that all of the switches are in the *closed* position, thus allowing current to pass from some point A to some point B. Suppose the order is given to cut the flow of current from A to B. In this case the failure of any of the switches to open will result in a failure, since current will continue to flow from A to B. A failure of a switch to open from a closed position is called a *short circuit failure* of the switch (or simply a *short*). Clearly, the more switches or diodes that are placed in parallel to control the flow of current between A and B, the greater is the probability that the parallel array of switches will fail short. Actually, if we want to protect ourselves against a short circuit type of failure, we would be better off placing the switches in a series arrangement, since now the flow of current from A to B can be cut off, if one or more of the switches obeys the command to go from a closed to an *open* position. However, the series arrangement is not panacea either, for suppose that the switches are all in an open position so no current flows from A to B. Suppose we now want current to flow between A and B and give the command to establish this objective. Clearly, the failure of any of the switches to close when ordered to do so will prevent current flowing from A to B. A failure of a switch to close when ordered to do so from an open position is called an *open circuit failure* (or simply an *open*).

How should the switches be arranged if we wish to reduce the probability of an open circuit type failure? The answer is to arrange the switches in parallel, for now current flows from A to B, if one or more of the switches obeys the command to go from an open position to a closed one. But this brings us full circle since, as we saw at the outset, a parallel arrangement of switches increases the probability of a short circuit type of failure. Consequently, if we wish protection against both short circuit and open circuit type failures, we need to employ parallel-series or series-parallel types of arrangements of switches. Just which arrangement we choose depends on which of the modes of failures is dominant (we assume tacitly that these are the only two ways in which a switch can fail and that these modes are independent and mutually exclusive).

Let us now compute the reliability of a series, parallel, series-parallel and parallel-series arrangements of switches.

3.3.1 Series arrangement

Consider first n switches in series $A_1, A_2, \ldots, A_n$. Let $F_i(t)$ be the time to failure c.d.f. of the ith switch, let p_i be the (conditional) probability that the switch fails open if a failure occurs and let $q_i = 1 - p_i$ be the (conditional) probability that the switch fails short. Since we are dealing with a series arrangement of switches, clearly the series array of switches will fail open on or before time t, if any of the switches fails open. This event has probability $1 - \Pi_{i=1}^{n}(1 - p_i F_i(t))$. Similarly, the series array of switches will fail short, if and only if, all of the switches fail short. This event has probability $\Pi_{i=1}^{n} q_i F_i(t)$. The two events are disjoint, thus the probability of the series array failing on or before time t is the sum of the two probabilities just obtained.

Hence, the reliability $R_s(t)$ of the series array of switches is given by

$$R_s(t) = \prod_{i=1}^{n}(1 - p_i F_i(t)) - \prod_{i=1}^{n} q_i F_i(t). \tag{3.28}$$

3.3.2 Parallel arrangement

Suppose that m switches are arranged in parallel $A_1, A_2, \ldots, A_m$. Let F_j, p_j and q_j be as before. Similar arguments lead us to $R_p(t)$, the reliability of the parallel array of switches,

$$R_p(t) = \prod_{j=1}^{m}(1 - q_j F_j(t)) - \prod_{j=1}^{m} p_j F_j(t). \tag{3.29}$$

3.3.3 Series-parallel arrangement

Suppose that nm switches A_{ij}, $i = 1, 2, \ldots, n$; $j = 1, 2, \ldots, m$ are in a series-parallel arrangement. By this we mean that n black boxes (figuratively speaking), each containing m switches in parallel, are arranged in series. Let $F_{ij}(t)$ be the time to failure c.d.f. of the jth switch in the ith black box and let p_{ij} and $q_{ij} = 1 - p_{ij}$ be, respectively, the (conditional) probability of an open or short of this switch, given that a failure of the switch has occurred. The reliability $R_{sp}(t)$ of this arrangement is given by

$$R_{sp}(t) = \prod_{i=1}^{n}\left(1 - \prod_{j=1}^{m} p_{ij} F_{ij}(t)\right) - \prod_{i=1}^{n}\left(1 - \prod_{j=1}^{m}(1 - q_{ij} F_{ij}(t))\right).$$
$$\tag{3.30}$$

3.3.4 Parallel-series arrangement

Suppose that mn switches A_{ji}, $j = 1, 2, \ldots, m$; $i = 1, 2, \ldots, n$ are in a parallel-series arrangement. By this we mean that m black boxes, each containing n switches in series, are arranged in parallel. Let F_{ji}, p_{ji} and q_{ji} be as before. Then, the reliability $R_{ps}(t)$ of this arrangement is given by

$$R_{ps}(t) = \prod_{j=1}^{m}\left(1 - \prod_{i=1}^{n} q_{ji} F_{ji}(t)\right) - \prod_{j=1}^{m}\left(1 - \prod_{i=1}^{n}(1 - p_{ji} F_{ji}(t))\right). \tag{3.31}$$

3.3.5 Simplifications

Formulae (3.28)–(3.31) become very much simpler, if all the switches have a common time to failure c.d.f. $F(t)$ and common conditional probability failing open, p, (hence of failing short $q = 1 - p$). If, as is often the case, we are interested in the reliability of various arrangements of switches for a specific length of time t^*, we define $p_o = pF(t^*)$ as the probability of a switch failing open in $(, t^*]$ and $p_s = qF(t^*)$ as the probability of a switch failing short in $(0, t^*]$. In practice, p_o and p_s are small and, in any case, $p_o + p_s = F(t^*) \leq 1$. Equations (3.28)–(3.31) become, respectively,

$$R_s(t^*) = (1 - p_o)^n - p_s^n, \tag{3.32}$$

$$R_p(t^*) = (1 - p_s)^m - p_o^m, \tag{3.33}$$

$$R_{sp}(t^*) = (1 - p_o^m)^n - (1 - (1 - p_s)^m)^n, \tag{3.34}$$

and

$$R_{ps}(t^*) = (1 - p_s^n)^m - (1 - (1 - p_o)^n)^m. \tag{3.35}$$

For a general reliability function R, define the *unreliability* function $\bar{R} = 1 - R$. With this notation, Equations (3.32)–(3.35) become

$$\bar{R}_s(t^*) = p_s^n + [1 - (1 - p_o)^n], \tag{3.36}$$

$$\bar{R}_p(t^*) = [1 - (1 - p_s)^m] + p_o^m, \tag{3.37}$$

$$\bar{R}_{sp}(t^*) = [1 - (1 - p_s)^m]^n + [1 - (1 - p_o^m)^n], \tag{3.38}$$

and

$$\bar{R}_{ps}(t^*) = [1 - (1 - p_s^n)^m] + [1 - (1 - p_o)^n]^m. \tag{3.39}$$

This representation is useful since it tells us precisely how much shorts and opens each contributes to the overall unreliability of various arrangements of switches. The reader should note the dualities in the above formulae. For example, for any arrangement of switches we can obtain the probabilities of failing open (short) from the probability of failing short (open) by replacing p_s (p_o) by $1 - p_o$ ($1 - p_s$) and subtracting the result from 1. Also R_p can be obtained from R_s (and vice versa) and R_{ps} from R_{sp} (and vice versa) by replacing p_o by p_s, p_s by p_o, m by n and n by m.

If $p_o = 0$, a switch can fail only as a short. In this case, the most preferable arrangement of a given number of switches to attain maximum reliability is a series arrangement. Similarly, if $p_s = 0$, a switch can fail only as an open. In this case, maximum reliability with a given number of switches is attained by a parallel arrangement.

For any choice of p_o and p_s, we can make the reliability arbitrarily close to 1 by appropriate choice of m and n.

For given p_o and p_s, it is possible to find n or m which maximize $R_s(t^*)$ or

$R_p(t^*)$. Maximization of expected system life, assuming a common exponential distribution is studied Barlow et al. [1].

3.3.6 Example

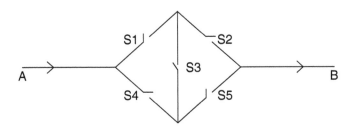

Figure 3.4 *A system with 5 switches.*

Consider the array of switches in Figure 3.4. Each switch is assumed to have probability p_o of failing open and p_s of failing short. Further assume that all the switches operate independently of each other. Suppose that all the switches are in the open position, so that no current can flow from A to B. Now, an order is given to close all the switches. Then, the probability that the array will fail open (i.e., current will not flow from A to B) is given by

$$1-2(1-p_o)^2-2(1-p_o)^3+5(1-p_o)^4-2(1-p_o)^5 = 2p_o^2+2p_o^3-5p_o^4+2p_o^5. \tag{3.40}$$

If all switches are in the closed position and the order is given to all switches to open, then the probability that the current flow will not be interrupted (short circuit failure) is given by

$$2p_s^2 + 2p_s^3 - 5p_s^4 + 2p_s^5. \tag{3.41}$$

Note the duality between the two last expressions.

It should be noted that the array in Figure 3.4 is neither series-parallel nor parallel-series. When all the switches are in the open position and the order is given for the switches to close, then current will in fact flow from A to B if and only if **both** S_1 **and** S_2 close, **or both** S_4 and S_5 close, **or** S_1, S_3, S_5 **all** close, **or** S_4, S_3, S_2 **all** close. Possible *paths* through which current can flow from A to B are $S_1 S_2$, $S_4 S_5$, $S_1 S_3 S_5$ and $S_4 S_3 S_2$. By the term *path* we mean a minimal set of switches whose correct functioning (in this case closing when ordered to do so from an open position) ensures correct functioning of the entire system.

Let A_1, A_2, A_3, A_4 be the events corresponding to paths $S_1 S_2$, $S_4 S_5$, $S_1 S_3 S_5$

and $S_4 S_3 S_2$ respectively closing when ordered to do so from an open position. Then $A_1 \cup A_2 \cup A_3 \cup A_4$ is the event that at least one path obeys the order to close and thus current can flow from A to B. The probability of this event is

$$P\{A_1 \cup A_2 \cup A_3 \cup A_4\} = P\{A_1\} + P\{A_2\} + P\{A_3\} + P\{A_4\}$$

$$-P\{A_1 A_2\} - P\{A_1 A_3\} - P\{A_1 A_4\} - P\{A_2 A_3\} - P\{A_2 A_4\} - P\{A_3 A_4\}$$

$$+P\{A_1 A_2 A_3\} + P\{A_1 A_2 A_4\} + P\{A_1 A_3 A_4\} + P\{A_2 A_3 A_4\} - P\{A_1 A_2 A_3 A_4\},$$

where

$$P\{A_1\} = P\{A_2\} = (1 - p_o)^2; \quad P\{A_3\} = P\{A_4\} = (1 - p_o)^3;$$

$$P\{A_1 A_2\} = P\{A_1 A_3\} = P\{A_1 A_4\} = P\{A_2 A_3\} = P\{A_2 A_4\} = (1 - p_o)^4;$$

$$P\{A_3 A_4\} = P\{A_1 A_2 A_3\} = P\{A_1 A_2 A_4\} = P\{A_1 A_3 A_4\}$$

$$= P\{A_2 A_3 A_4\} = P\{A_1 A_2 A_3 A_4\} = (1 - p_o)^5.$$

Denoting by $\bar{C}$ the complement of any event C, then $\overline{A_1 \cup A_2 \cup A_3 \cup A_4} = \bar{A}_1 \bar{A}_2 \bar{A}_3 \bar{A}_4$ is the event that current does **not** flow through any of the paths when the order is given for the switches to close from an open position (i.e., the system in Figure 3.4 fails open). The probability of failing open is

$$P\{\bar{A}_1 \bar{A}_2 \bar{A}_3 \bar{A}_4\} = 1 - P\{A_1 \cup A_2 \cup A_3 \cup A_4\}.$$

The probability that this system fails open can be calculated in another way. When all switches are open and the order is given for the switches to close, the current will in fact *not* flow from A to B if **both** S_1, S_4 remain open, or **both** S_2, S_5 remain open, or S_1, S_3, S_5 **all** remain open, or S_1, S_3, S_5 **all** remain open. The segments $S_1 S_4$, $S_2 S_5$, $S_1 S_3 S_5$ and $S_4 S_3 S_2$ are called *cuts* because they represent different ways in which current flow from A to B can be prevented. By the term *cut* we mean a minimal set of switches whose incorrect functioning (in this case remaining open when ordered to close) ensures failure of the entire system.

Let B_1, B_2, B_3, B_4 be the events corresponding to cuts $S_1 S_4$, $S_2 S_5$, $S_1 S_3 S_5$, $S_4 S_3 S_2$ respectively remaining open when ordered to close. Then, $B_1 \cup B_2 \cup B_3 \cup B_4$ is the event that current flow from A to B is prevented by at least one of the cuts when the order is given for the switches to close from an open position. The probability of this event is

$$P\{B_1 \cup B_2 \cup B_3 \cup B_4\} = P\{B_1\} + P\{B_2\} + P\{B_3\} + P\{B_4\}$$

$$-P\{B_1 B_2\} - P\{B_1 B_3\} - P\{B_1 B_4\} - P\{B_2 B_3\} - P\{B_2 B_4\} - P\{B_3 B_4\}$$

$$+P\{B_1 B_2 B_3\} + P\{B_1 B_2 B_4\} + P\{B_1 B_3 B_4\} + P\{B_2 B_3 B_4\} - P\{B_1 B_2 B_3 B_4\},$$

where

$$P\{B_1\} = P\{B_2\} = p_o^2; \quad P\{B_3\} = P\{B_4\} = p_o^3;$$

$$P\{B_1 B_2\} = P\{B_1 B_3\} = P\{B_1 B_4\} = P\{B_2 B_3\} = P\{B_2 B_4\} = p_o^4;$$

$$P\{B_3B_4\} = P\{B_1B_2B_3\} = P\{B_1B_2B_4\} = P\{A_1B_3B_4\}$$
$$= P\{B_2B_3B_4\} = P\{B_1B_2B_3B_4\} = p_o^5.$$

It is readily verified that $P\{\bar{A}_1\bar{A}_2\bar{A}_3\bar{A}_4\} = P\{B_1\cup B_2\cup B_3\cup B_4\}$ is the result given in Equation (3.40). In a completely analogous way, one can compute the probability that the system fails short. The details are left to the reader.

By using paths and cuts it is possible, in principle, to compute exactly the reliability of any arrangement of components given the reliability of the individual components (under complete independence). However, the exact calculations may involve so much effort that it becomes impractical to carry them out. It is, therefore, of interest that for a useful class of arrangements which are called *coherent* or *monotonic* (these include the arrangements discussed in this chapter as special cases) it is possible to find upper and lower bounds for the reliability. This is done using the paths and cuts method. Roughly speaking, this involves approximating a complex arrangement of components by a series-parallel and parallel-series arrangements. For more details and examples see Barlow and Proschan [3].

3.4 Standby redundancy

A common method for improving reliability is to have a supply of spares available to replace a unit when it fails. Spares are used one at a time as replacements and are assumed to be "inactive" (i.e., not aging while in standby status). It is further assumed that the replacement time required to put a spare in place of a failed unit is zero and that switching, if required in making the replacement, is perfect. The original and standby units have a common time of failure c.d.f. $F(t)$. Failure is said to occur when the original unit and all the standby spares have failed. If $T_1, T_2, \ldots, T_n$ are the life times of the original unit and the $(n-1)$ spares, then the time of system failure Z_n is given by

$$Z_n = T_1 + T_2 + \cdots + T_n. \tag{3.42}$$

The c.d.f. $F_n(t)$ of Z_n can be found by various methods, e.g., repeated use of convolution formulae, or by the use of moments generating or characteristic functions. Since the T_i are assumed to be independent random variables with common c.d.f. $F(t)$, we can assert that

$$E(Z_n) = nE(T_n) = n\mu \tag{3.43}$$

and

$$\mathrm{Var}\, Z_n = n\mathrm{Var}\, T_i = n\sigma^2, \tag{3.44}$$

where μ and σ^2 are the mean and variance of the time of failure for a single unit. In those cases where n is sufficiently large, it will follow from the central limit theorem that Z_n is approximately normally distributed with mean $n\mu$ and variance $n\sigma^2$ (written as $Z_n \sim N(n\mu, n\sigma^2)$). The symbol $\sim$ stands for "is distributed as."

There are some important cases where the distribution of Z_n can be found exactly. Among these are:

(a) The T_i are normally distributed with common mean μ and common variance σ^2, i.e., $T_i \sim N(\mu, \sigma^2)$. In this case, $Z_n = T_1 + T_2 + \cdots + T_n \sim N(n\mu, n\sigma^2)$. System reliability is given by

$$R_n(t) = P(Z_n > t) = 1 - \Phi\left(\frac{t - n\mu}{\sigma\sqrt{n}}\right), \tag{3.45}$$

where $\Phi(x) = (2\pi)^{-1/2} \int_{-\infty}^{x} e^{-u^2/2} du$ is the c.d.f. of a standard normal random variable.

(b) The T_i are exponentially distributed with common probability density function $f(t) = \lambda e^{-\lambda t}$, $t \geq 0$. Since the exponential distribution can be considered as the distribution of time intervals between successive events in a Poisson process with event rate λ, it follows that $Z_n = T_1 + T_2 + \cdots + T_n$ can be considered as the time of the n'th event in a homogenous Poisson process with event rate λ. In Section 1.2 we saw that the p.d.f. of Z_n is given by the gamma (or Erlang) distribution

$$f_n(t) = \lambda(\lambda t)^{n-1} \frac{e^{-\lambda t}}{(n-1)!}, \quad t \geq 0. \tag{3.46}$$

The system reliability is given by

$$R_n(t) = \sum_{j=0}^{n-1} e^{-\lambda t} \frac{(\lambda t)^j}{j!}. \tag{3.47}$$

This result can be obtained formally as $\int_t^\infty f_n(\tau) d\tau$. A much more direct proof is to note that the original unit and its replacement generate a Poisson process having rate λ. The poisson process does not go on indefinitely but terminates when the last replacement fails. Hence $R_n(t)$ is simply the probability that for this Poisson process at most $(n-1)$ events occur in $(0, t]$. This is precisely Equation (3.47).

3.5 Problems and comments

Problems for Section 3.1

1. Verify that $G_s(t)$ in Equation (3.5) is the distribution of the smallest observation in a random sample of size n drawn from an underlying c.d.f. $F(t)$.

2. Verify that the smallest value in a sample of size n drawn at random from an exponential distribution with mean θ is also exponential with mean θ/n.

3. A sample size n is drawn at random from a Weibull distribution described by the c.d.f. $F(t) = 1 - e^{-\alpha t^\beta}$; $t \geq 0$, $\beta > 0$. Prove that the smallest value in the sample of size n follows the Weibull distribution $G(t) = 1 - e^{-n\alpha t^\beta}$, $t \geq 0$.

4. Suppose that we have a system composed of a series of n units, where the c.d.f. of the time to failure of the ith unit is given by the Weibull distribution, $F_i(t) = 1 - e^{-\alpha_i t^\beta}$; $\alpha_i, \beta > 0$; $t \geq 0$. Find $F(t)$, the reliability of the system.

5. Let $h_i(t)$ be the hazard function of the ith subsystem in a system composed of n subsystems in series. Prove that the system hazard function $h_s(t) = \sum_{i=1}^{n} h_i(t)$.

6. Verify that for values of t for which $F(t)$ is "very small," $G_s(t) = 1 - [1 - F(t)]^n \simeq nF(t)$. Check this approximation numerically when $F(t) = .01$ and $n = 5$.

7. Show that if the $F_i(t)$, $i = 1, 2, \dots, n$ are each IFR (DFR), then $G_s(t)$ is also IFR (DFR).

8. Consider the successive failure times of a series system composed of two independent subsystems A and B. For simplicity, replacement of a failed unit is assumed to be instantaneous. Suppose that the failure times of A are exponentially distributed with mean $\theta_A = 100$ hours and the failure times of B are exponentially distributed with mean $\theta_B = 200$ hours.

   (a) What is the probability that exactly three A failures occur between two successive B failures?

   (b) What is the probability that at least three A failures occur between two successive B failures?

   (c) What is the probability that exactly three A failures occur during a 200-hour interval?

   (d) What is the probability that at least three A failures occur during a 200-hour interval?

9. Consider a device consisting of n parts $A_1, A_2, \dots, A_n$. The device is subjected from time to time to shocks, which affect all parts simultaneously. The shocks are assumed to occur according to a Poisson process with rate λ. It is assumed that individual parts will fail or survive independently of the fate of any other parts, with part A_i having conditional probability $p_i(q_i = 1 - p_i)$ of failing (surviving), given that a shock occurs. Failure of any part causes failure of the device.

   (a) Show that $R(t)$, the probability that the device survives for a length of time t, is given by
   $$e^{-\lambda\{1 - \prod_{i=1}^{n} q_i\}t}.$$

   (b) Show that $1 - \prod_{i=1}^{n} q_i \leq \sum_{i=1}^{n} p_i$ and hence that $R(t) \geq e^{-\lambda \sum_{i=1}^{n} p_i t}$, the estimate of system reliability corresponding to adding part failure rates, where part A_i has failure rate λp_i.

(c) Suppose $p_1 = p_2 = \cdots = p_n = p$. Find numerical values for $e^{-\lambda\{1-q^n\}t}$ and $e^{-n\lambda pt}$ for $\lambda = .0001/\text{hour}$, $p = .01$, $n = 10,000$, $t = 100$ hours.

10. Let us assume that the life-time of a certain part is exponentially distributed with p.d.f. $\theta^{-1}e^{-t/\theta}$, $t \geq 0$. Suppose that a sample of these parts has been placed on life test for a total of T part-hours. During this time, $N(T)$ part-failures were observed. Consider a system consisting of n of these parts in series. Failure of any part causes the system to fail. Verify using the results in Chapter 1, that an unbiased estimate of system reliability in the interval $(0, t^*]$, with $t^* < T/n$ is given by $(1 - nt^*/T)^{N(T)}$.

Hint: System reliability in $(0, t^*] = e^{-nt^*/\theta}$.

11. Consider a system S with n independent subsystems in series $S_1, S_2, \ldots, S_n$. The life-time of subsystem S_i is exponentially distributed with mean life θ_i. Suppose that subsystem S_i is placed on test for a length of time T_i and the observed (random) number of failures is K_i (it is assumed that when a subsystem fails it is repaired immediately). Verify that an unbiased estimator of system reliability in the interval $(0, t^*]$, with $t^* < \min\{T_i\}$, is given by

$$\prod_{i=1}^{n} \left(1 - \frac{t^*}{T_i}\right)^{K_i}.$$

Note, if $T_1 = T_2 = \cdots = T_n = T$, this estimator becomes

$$\left(1 - \frac{t^*}{T}\right)^{\Sigma K_i}.$$

12. Reconsider the system in Problem 10. Suppose that instead of life-testing for a total of T part-hours, life-testing on a sample of parts is carried out until the rth failure occurs with associated (random) total life T_r. Show that

$$\mathbf{1}\{T_r \geq nt^*\} \cdot \left(1 - \frac{nt^*}{T_r}\right)^{r-1}$$

is unbiased for system reliability over $(0, t^*]$.

13. Reconsider the system in Problem 11. Suppose that subsystem S_i placed on life-test until the rth failure occurs and this happens at total life $T_{r,i}$. Show that

$$\mathbf{1}\{\min_i T_{r,i} \geq t^*\} \cdot \prod_{i=1}^{n} \left(1 - \frac{t^*}{T_{r,i}}\right)^{r-1}$$

is unbiased for system reliability over $(0, t^*]$.

Comment. The estimators in Problems 10–13 are all UMVUE.

14. Prove that $G_p(t)$ in Equation (3.8) is the distribution of the largest observation in a random sample of size n drawn from an underlying c.d.f. $F(t)$.

15. Prove Equation (3.10).

16. Compare graphically the system reliability of a two unit parallel redundant system (each unit is assumed to fail exponentially with mean θ) and a single unit having mean $3\theta/2$. In drawing the graph use the dimensionless scale, t/θ, as the abscissa.

17. A unit has a life time which is exponentially distributed with mean $\theta = 1,000$ hours. How many units should be placed in service simultaneously to ensure that at least one unit survives for a time $t = 1,000$ hours with probability $\geq .99$?

18. Consider a redundant system consisting of three units A_1, A_2, A_3 in parallel. Unit A_i has life time p.d.f. $\theta_i^{-1}e^{-t/\theta_i}$, $t \geq 0$, $i = 1, 2, 3$. Find $R(t)$, the reliability of the system. What is the probability that the system survives 100 hours if $\theta_1 = 100$, $\theta_2 = 2,000$, $\theta_3 = 300$ hours? What is the mean time of system failure?

19. A redundant system consists of three units B_1, B_2, B_3 in parallel. The life time of each unit is distributed with p.d.f. $f(t) = t^2e^{-t}/2$, $t \geq 0$. What is the probability that the system survives for a length of time $t^* = 1$?

20. A redundant system consists of two units A_1 and A_2 in parallel. The life time of A_1 is exponentially distributed with p.d.f. $f_1(t) = e^{-t}$, $t \geq 0$ and the life time of A_2 is exponentially distributed with p.d.f. $f_2(t) = 2e^{-2t}$, $t \geq 0$.

(a) Find the c.d.f. of system life.

(b) Find the system hazard rate.

(c) Show that the hazard rate is not monotone and compute the maximal value of the hazard rate.

Comment. It can be shown that the system life follows an IHRA distribution. In this problem we have an example of a life distribution which is IHRA but not IFR.

21. A system S consists of n units in parallel redundancy, each having common life time c.d.f. $F(t)$. Show that if $F(t)$ is IFR, then the c.d.f. of the life time of S is also IFR.

22. A system consists of two units, A and B in parallel redundancy. Unit A is assumed to have constant failure rate α and unit B is assumed to have constant failure rate β as long as neither A nor B has failed. If, however, A fails before B then the failure rate of B changes to β'. Similarly, if B fails before A, then the failure rate of A changes to α'.

(a) If the random variable X is the time of failure of A and the random variable Y is the time of failure of B, find $f(x, y)$, the joint p.d.f. of the failure times of A and B.

(b) Find $f(x)$, the p.d.f. of the time of failure of unit A and $g(y)$, the p.d.f. of the time of failure of unit B.

(c) Find $E(X)$, VarX, $E(Y)$, VarY, $\text{Cov}(X, Y)$ and
$\rho(X, Y) = \text{Cov}(X, Y)/\sigma_x \sigma_Y$.

(d) Suppose that system failure occurs if and only if both units A and B fail. Find $R(t)$, the probability that the system survives time t.

23. A system consists of two units A and B in parallel redundancy. The system is subjected from time to time to "shocks" arising from a Poisson process with rate λ. When a shock occurs, each of the units has conditional probability of failing, p and of surviving, $q = 1 - p$. It is also assumed that a unit has no memory, i.e., if it has survived $(k - 1)$ shocks then the conditional probability of failing when the k'th shock occurs remains equal to p. System failure occurs by time t, if and only if both units A and B have failed by time t.

(a) Show that $R(t)$, the probability that the system does not fail in the interval $(0, t]$ is equal to $2e^{-\lambda pt} - e^{-\lambda(1-q^2)t}$.

(b) Find the mean time to system failure and the variance of the time to system failure.

(c) What is the probability that the system will fail because both units A and B fail simultaneously in $(, t]$?

(d) What is the probability that exactly j units survive for a length of time t, $j = 0, 1, 2$?

24. (Continuation of Problem 23). Suppose that the system consists of n units in parallel redundancy and that otherwise the model is the same as in Problem 23.

(a) Show that the reliability for the system is given by

$$R(t) = \sum_{j=1}^{n} (-1)^{j+1} \binom{n}{j} e^{-\lambda(1-q^j)t}.$$

(b) Show that $P_{[j]}(t)$, the probability that **exactly** j units survive for a length of time t is given by

$$P_{[j]}(t) = \binom{n}{j} \sum_{i=0}^{n-j} (-1)^i \binom{n-j}{i} e^{-\lambda(1-q^{i+j})t}.$$

25. Consider a system composed of two units A_1 and A_2. Suppose that shocks to A_1 alone occur according to a Poisson process with rate λ_1, shocks to A_2 alone occur according to a Poisson process with rate λ_2, and shocks to both A_1 and A_2 occur with rate λ_{12}. It is assumed that when a shock occurs the

affected unit (or units) will fail. Let the random variable $X(Y)$ denote the life of unit $A_1(A_2)$.

(a) Show that $P(X > s, Y > t) = \exp\{-\lambda_1 s - \lambda_2 t - \lambda_{12}\max(s,t)\}$.

(b) Show that f(x,y), the joint p.d.f. of X and Y is given by

$$f(x,y) = \mathbf{1}\{x < y\} \cdot \lambda_1(\lambda_2 + \lambda_{12})e^{-\lambda_1 x - (\lambda_2 + \lambda_{12})y}$$

$$+ \mathbf{1}\{x > y\} \cdot \lambda_2(\lambda_1 + \lambda_{12})e^{-(\lambda_1 + \lambda_{12})x - \lambda_2 y}$$

$$+ \mathbf{1}\{x = y\} \cdot \lambda_{12}e^{-\lambda x},$$

where $\lambda = \lambda_1 + \lambda_2 + \lambda_{12}$.

(c) Show that f(x), the (marginal) p.d.f. of X, is $(\lambda_1 + \lambda_{12})e^{-(\lambda_1 + \lambda_{12})x}$, $x \geq 0$; and that g(x), the (marginal) p.d.f. of Y, is $(\lambda_2 + \lambda_{12})e^{-(\lambda_2 + \lambda_{12})y}$, $y \geq 0$.

26.(a) (Continuation of Problem 25.) Suppose that the two units A and B are in series so that a failure of either unit causes system failure. Show that the system reliability is $R_s(t) = e^{-\lambda t}$, where $\lambda = \lambda_1 + \lambda_2 + \lambda_{12}$.

(b) Suppose that the two units A and B are in parallel so that system failure occurs only when both units are failed. Show that the system reliability is

$$R_p(t) = e^{-(\lambda_2 + \lambda_{12})t} + e^{-(\lambda_1 + \lambda_{12})t} - e^{-\lambda t}.$$

27. Let us assume that the life time of a certain part is exponentially distributed with mean θ. Suppose that a sample of these parts has been placed on life test for a total of T part-hours. During this time $N(T)$ part failures were observed. Consider a system consisting of two of these parts in parallel. System failure occurs if and only if both parts fail. Verify using the results of Chapter 1, that an unbiased estimate of system reliability in interval $(0, t^*]$, with $t^* < T/2$, is given by

$$2\left(1 - \frac{t^*}{T}\right)^{N(T)} - \left(1 - \frac{2t^*}{T}\right)^{N(T)}.$$

Hint: System reliability in $(0, t^*] = 2e^{-t^*/\theta} - e^{-2t^*/\theta}$.

28. (Continuation of Problem 27). Suppose that the system consists of n parts in parallel redundancy and that system failure occurs if and only if all n parts fail. Verify using the results of Chapter 1, that an unbiased estimate of system reliability in the interval $(0, t^*]$, with $t^* < T/n$, is given by

$$\sum_{k=1}^{n}(-1)^{k-1}\binom{n}{k}\left(1 - \frac{kt^*}{T}\right)^{N(T)}.$$

29. (Continuation of Problem 28). Suppose for the system in Problem 28 that instead of testing parts for a total of T part-hours, a sample of parts are

tested until the r'th failure occurs with associated total life T_r. Verify that

$$\sum_{k=1}^{n}(-1)^{k-1}\binom{n}{k}\left(1-\frac{kt^*}{T_r}\right)_+^{r-1}$$

is an unbiased estimate of system reliability.

Comment. The unbiased estimates in Problems 27–29 are actually UMVUE.

30. Suppose that the reliability of an airplane engine for a ten-hour flight is .99. At least two out of the four engines must operate properly for the entire ten-hour period in order to have a successful flight. Compute the probability of this event, assuming that the engines operate independently and that failure of one or two engines does not effect the probability of failure of the other engines.

31. Three elements in series have reliabilities of .9, .8, .7 for some specified time period. Find the reliability of the system composed of these elements.

32. Suppose that an item has probability .8 of surviving ten hours. Suppose that three of the items are placed in a missile. What is the probability that at least one of them survives ten hours? What is the probability that all three items survive ten hours? What is the probability that at least two of the items survive ten hours?

33. A certain operation can be performed by either one of two systems A and B. Assume that the systems A and B operate completely independently of each other and that the probability of A functioning properly is .8 and of B functioning properly is .7. Compute the probability that the operation is performed successfully by at least one of the two systems.

34. A system is composed of three units, each of which has reliability .9 for some desired time of operation. Assuming that the system works, if at least one of the units works, find the system reliability.

35. Ten satellites each having life time p.d.f. $f(t) = .5e^{-t/2}, t \geq 0$ are launched at the same time. What is the probability that at least five of the satellites will still be functioning properly one year after the launch?

36. Show that the life time c.d.f. of a k out of n system composed of units having common IFR life time c.d.f., $F(t)$, is also IFR.

37. A device consists of three parts A_1, A_2 and A_3. The device is subjected from time to time to "shocks" arising from a Poisson process with rate λ. When a shock occurs, each of the parts has conditional probability p of failing and $q = 1 - p$ of not failing. The device works properly if at least two of the parts work. Show that $R(t)$, the probability that the device survives for a length of time t, is given by

$$R(t) = 3e^{-\lambda\left[1-q^2\right]t} - 2e^{-\lambda\left[1-q^3\right]t}.$$

Show that for p "small"

$$R(t) \approx 3e^{-2\lambda pt} - 2e^{-3\lambda pt}.$$

38. (Extension of Problem 37). A device consists of n parts $A_1, A_2, \ldots, A_n$. The device is subjected from time to time to "shocks" arising from a Poisson process with rate λ. When a shock occurs, each of the parts has conditional probability p of failing and $q = 1 - p$ of not failing. The device works if at least k out of n of the parts work. Show that $R_{k,n}(t)$, the probability that the device survives for a length of time t, is given by

$$R_{k,n}(t) = \sum_{j=k}^{n} P_{[j]}(t),$$

where

$$P_{[j]}(t) = \binom{n}{j} \sum_{i=0}^{n-j} (-1)^i \binom{n-j}{i} e^{-\lambda(1-q^{j+i})t}, \quad k \le j \le n.$$

Problems for Section 3.2

1. Consider a series-parallel system with $n = 4$ and $m_1 = m_2 = m_3 = m_4 = m$. Suppose that each of the mn components in the system has common c.d.f. $F(t) = 1 - e^{-t/100}$, $t \ge 0$. What is the minimum m for which the system reliability for $t^* = 10$ hours $\ge .99$?

2. Consider a parallel-series system with $n_1 = n_2 = n_3 = n_4 = n = 4$. Suppose that each of the mn components in the system has common c.d.f. $F(t) = 1 - e^{-t/100}$, $t \ge 0$. What is the minimum m for which the system reliability for $t^* = 10$ hours $\ge .99$?

3. Consider a series-parallel system consisting of subsystems A_1, A_2 and A_3 in series. A_1 consists of two elements A_{11} and A_{12} in parallel having common c.d.f. $1 - e^{-t/10}$, $t \ge 0$; A_2 consists of a single element with c.d.f. $1 - e^{-t/100}$, $t \ge 0$; A_3 consists of three elements A_{31}, A_{32}, A_{33} in parallel having common c.d.f. $1 - e^{-t/5}$, $t \ge 0$.

   (a) Find $R_{sp}(t)$, the reliability of the system, and compute $R_{sp}(t^*)$ for $t^* = 1$.

   (b) Find the MTBF of the system.

   (c) Compute the reliability of the system, whose time of failure is exponentially distributed with mean equal to the MTBF computed in (b) and plot this reliability as a function of t on the same graph as $R_{sp}(t)$.

4. (a) For $m = n = 2$, prove Equation (3.27).

   (b) For $m = 3$, $n = 2$, assuming all units have the same reliability r, prove that

$$r_{sp} - r_{ps} = 6r^2(1 - r)^3,$$

where r_{sp} and r_{ps} are given by Equations (3.21) and (3.26), respectively.

Problems for Section 3.3

1. Suppose that $p_o = .1$ and $p_s = .05$. What are the values of n and m which maximize R_s and R_p, respectively?

2. Suppose that $p_o + p_s = \alpha$ with $p_o = k\alpha$, and $p_s = (1 - k)\alpha$, where $0 \leq k \leq 1$ and $0 \leq \alpha \leq 1$. Show that if $k < 1/2$ ($k > 1/2$) a series (parallel) arrangement of two identical switches is more reliable than a single switch. What happens if $k = 1/2$?

3. (a) Show that if $p_o = p_s = \alpha/2$, $n = m = 2$, then
$$\bar{R}_{sp}(t^*) = \bar{R}_{ps}(t^*) = 1.5\alpha^2 - .5\alpha^3.$$

   (b) Show under the assumption in (a), that for a series-parallel arrangement of four switches the probability of failing short exceeds the probability of failing open and that the reverse is true for parallel-series arrangement.

4. (a) Show that the probability that four identical switches will fail short (open) when arranged series-parallel is less than or equal to the probability that a single switch will fail short (open) provided that p_s (p_o) for a single switch is $< .382$ ($> .618$).

   (b) Give similar results for the case where four identical switches are in parallel-series array.

5. (a) Show that if a switch can only fail as short, then for the case $m = n = 2$, $\bar{R}_{sp}(t^*) \geq \bar{R}_{ps}(t^*)$.

   (b) Show that if a switch can only fail as open, then for the case $m = n = 2$, $\bar{R}_{sp}(t^*) \leq \bar{R}_{ps}(t^*)$.

6. Consider the array of switches in Figure 3.5. Each switch is assumed to have probability p_o of failing open and p_s of failing short and it is assumed that all switches operate independently of each other.

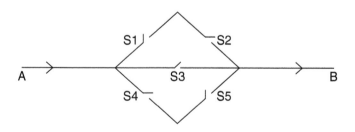

Figure 3.5 *A system with 5 switches, Problem 6.*

(a) Suppose that all switches are in the open position, so that no current flows from A to B. Show that if the order is given to close the switches, then the probability that current will **not** flow from A to B (or equivalently, the probability that the array of switches will fail open) is given by

$$1 - (1 - p_o) - 2(1 - p_o)^2 + 2(1 - p_o)^3$$
$$+ (1 - p_o)^4 - (1 - p_o)^5 = p_o^3 (2 - p_o)^2.$$

(b) Suppose that all switches are in the closed position and that the order is given to interrupt the flow of current from A to B. Show that the probability that the current flow will **not** be interrupted (or equivalently, the probability that the array of switches will fail short) is given by

$$p_s + 2p_s^2 - 2p_s^3 - p_s^4 + p_s^5.$$

(c) How are (a) and (b) related?

Problems for Section 3.4

1. Compare graphically the system reliability of a unit plus two inactive spares, where each unit is assumed to have a life time which is exponentially distributed with mean θ, with the reliability of a single unit having a life time which is exponentially distributed with mean 3θ.

2. Suppose that a battery has a life time which is normally distributed with $\mu = 100$ hours, and $\sigma = 10$ hours. Assume that a battery, not in use, does not age. What is the probability that a battery and 3 inactive spares will last longer than 350 hours?

3. A unit has a life time which is exponentially distributed with mean $\theta = 1,000$ hours. How many inactive spares are needed to ensure that at least one unit (original or spares) survives for a time $t = 1,000$ hours with probability .99?

4. Suppose that the life time of a unit is distributed with p.d.f. $f(t) = \lambda(\lambda t)^{a-1} e^{-\lambda t}/\Gamma(a)$, $t \geq 0$, $a > 0$. Prove that Z_n, the total life time of the original unit plus $(n-1)$ inactive spares each having the above life distribution, is distributed with p.d.f. $f_n(t) = \lambda(\lambda t)^{na-1} e^{-\lambda t}/\Gamma(na)$, $t \geq 0$.

5. We are given a unit and a partially active spare. It is assumed that the unit has a constant failure rate λ_1. During the time that the spare is in standby status, it has a constant failure rate $0 \leq \lambda_2 \leq \lambda_1$. When the original unit fails and the spare (if available) is switched on, its failure rate raises from λ_2 to λ_1. Find $R(t)$, the probability that at least one of these units survives for a length of time t.

6. A piece of radar equipment on an airplane is assumed to have a constant failure rate $\lambda = .01/\text{hour}$.

(a) What is the probability that the radar equipment will fail in a ten-hour flight?

(b) If an identical piece of radar equipment is carried as insurance, compute the probability that at least one of the radars is in working order at the end of the flight, if both radar units are in use simultaneously throughout the flight.

(c) Same as (b) except the second unit is an "inactive" standby spare.

It is assumed that repair of the radar is not possible during the flight.

7. Consider a system consisting of an active unit and two partially active standby units. The active unit is assumed to have a constant failure rate λ_1 and each standby unit has a constant failure rate λ_2, with $0 \leq \lambda_2 \leq \lambda_1$. If the active unit should fail, one of the standby units would be switched on immediately (perfect switching assumed) with failure rate rising from λ_2 to λ_1. The second standby unit would similarly be switched on (with failure rate rising from λ_2 to λ_1) if required by a second failure.

(a) Assuming that failed units cannot be repaired, find $R(t)$, the probability that at least one of the three units will survive for a length of time t.

(b) The system will be said to fail as soon as all three units have failed. Find the expected time until system failure occurs.

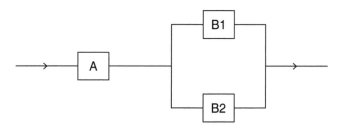

Figure 3.6 *Block diagram for Problem 8.*

8. Consider the system represented by the block diagram in Figure 3.6. For the system to function successfully subsystem A and at least one of the two subsystems B_1 and B_2 must function successfully. Assuming that A has a life time which is distributed with p.d.f. $\lambda_1^2 t e^{-\lambda_1 t}$, $t \geq 0$ and that B_1 and B_2 have lives distributed with p.d.f. $\lambda_2 e^{-\lambda_2 t}$, $t \geq 0$, find the reliability of the system and the mean time to system failure if:

(a) B_1 and B_2 are simultaneously in service.

(b) B_2 is in standby redundancy to B_1, i.e., has failure rate zero unless B_1 fails. When this happens the failure rate of B_2 becomes λ_2.

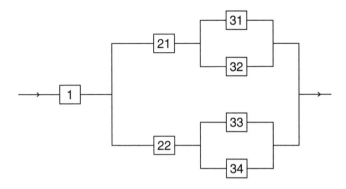

Figure 3.7 *Block diagram for Problem 9.*

9. Consider the system represented by the block diagram in Figure 3.7. The following assumptions are made:

   (a) The branch containing units 22, 33 and 34 is inactive and has failure rate zero until the failure of the brunch containing units 21, 31 and 32.

   (b) Unit 32 is in standby redundancy to unit 31. Unit 32 has failure rate zero until the failure of unit 31.

   (c) Unit 34 is in standby redundancy to unit 33. Unit 34 has failure rate zero until the failure of unit 33.

   (d) Unit 1 is continuously active.

   (e) Unit 1 has constant failure rate λ_1, unit 21 has constant failure rate λ_2, and unit 22 has also failure rate λ_2 when it becomes active; unit 31 has constant failure rate λ_3 and so do units 32, 33 and 34 when they become active.

   Find $R(t)$, the reliability of the system.

10. Consider the two-stage system represented by the block diagram in Figure 3.8. Units A and D are initially active; B and C are inactive standby units for A; E is an inactive standby unit for D. If A fails, B is activated. If D fails, both B and E are activated. Units C and E are activated, if B fails. Assuming that each unit has life to failure p.d.f. $\lambda e^{-\lambda t}$, $t \geq 0$ when active (and failure rate zero when inactive), find $R(t)$, the reliability of the system. Also find the mean time to system failure. If $\lambda = .01$, find $R(t)$ for $t = 20$ hours and ET.

11. Same as Problem 10, except that all units operate simultaneously.

12. A system consists of n active units in parallel. It is assumed that each unit has constant failure rate λ/n. When a unit fails, the failure rate of the re-maining $(n-1)$ good units becomes $\lambda/(n-1)$. When j units have failed,

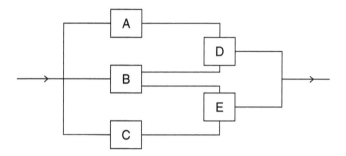

Figure 3.8 *Block diagram for Problem 10.*

the remaining $(n-j)$ good units each has failure rate $\lambda/(n-j)$. The system is said to fail when all units have failed. Find the p.d.f. of the time to system failure and the mean time to system failure. If $n = 4$ and $\lambda = .1/$hours, find the probability that the system will survive 20 hours.

13. We are given three nonrepairable systems: A, B and C. System A consists of one unit with life which is exponentially distributed with rate λ. System B is a parallel system with n independent units, each one has exponential life with same rate λ. System C consists of n such units, but only one is active and $n - 1$ units in full redundancy.

   (a) What is the probability that system A fails before system B?

   (b) What is the probability that system A fails before system C?

   (c) What is the probability that system B fails before system C?

14. A system consists of n identical units in series, each unit having time to failure p.d.f. $f(t) = \lambda e^{-\lambda t}$, $t \geq 0$. In addition, k inactive spares are available and are switched on automatically if needed (perfect switching assumed), thus keeping the system running. System failure occurs when no spare is available to replace a failed unit. Find the p.d.f. of the system life T and the mean time to system failure.

15. Let us assume that the life time of a certain part is exponentially distributed with p.d.f. $\lambda e^{-\lambda t}$, $t \geq 0$. Suppose that a sample of these parts is placed on life test for a total of T part-hours. During this time $N(T)$ part failures were observed. Suppose that at $t = 0$ we place one of these items in service and that we have $(n - 1)$ inactive spares available as replacements as item failures occur. Verify that an unbiased estimate of $R(t^*) = e^{-\lambda t^*} \sum_{j=0}^{n-1} (\lambda t^*)^j / j!$, the probability that a spare is available as a replace-

ment when needed in $(0, t^*]$, $t^* < T$, is given by

$$\sum_{j=0}^{n-1} \binom{N(T)}{j} \left(\frac{t^*}{T}\right)^j \left(1 - \frac{t^*}{T}\right)^{N(T)-j}.$$

16. Suppose that we have a single unit and two inactive spares with underlying p.d.f. $\lambda e^{-\lambda t}$, $t \geq 0$.

   (a) Suppose that three failures were observed in one thousand unit-hours of life testing of a sample of these units. Estimate the probability that we will not run out of spares in 100 hours.

   (b) Estimate the probability of not running out of spares in 100 hours if there had been four failures in $1,000$ hours; five failures in $1,000$ hours.

17. Suppose in the previous problem that a life test was carried out on a sample of these units until the fifth failure occurs. Suppose that T_5, the total life up to and including the fifth failure, was 1,000 unit-hours. Estimate the probability of not running out of spares in 100 hours.

18. A unit has a life time which is assumed to be exponentially distributed with p.d.f. $f(t) = \lambda e^{-\lambda t}$, $t \geq 0$. If the unit fails, a standby inactive spare unit having the same life time p.d.f. is placed into service without delay. Show that an excellent approximation to the probability that both, the unit and the spare, fail by time t $(\lambda t \leq .1)$ is given by $(\lambda t)^2/2 - (\lambda t)^3/3$ for $\lambda t < .1$.

CHAPTER 4

Reliability of a one-unit repairable system

4.1 Exponential times to failure and repair

In this chapter we consider stochastic models in which a failed unit can be repaired. The easiest problem to treat is the special but important case where we have a single unit, which is repaired as soon as it fails and put into service without delay as soon as the repair has restored it to an operating or good state. We assume that the time to failure distribution is described by the exponential p.d.f. $f(t) = \lambda e^{-\lambda t}$, $t \geq 0$ and that the time to repair distribution is described by the exponential p.d.f. $g(t) = \mu e^{-\mu t}$, $t \geq 0$. The rates λ and μ are known, respectively, as the *system failure rate* and *system repair rate*. The mean time to failure (MTF) and mean time to repair (MTR) are, respectively, $1/\lambda$ and $1/\mu$.

It is convenient to think of the unit as being in one of two possible states:

$$E_1 \text{ (good state);} \quad E_0 \text{ (bad state).}$$

Assuming exponentially distributed times to failure and repair means, in particular, that we have a stochastic process with (conditional) transition probabilities in the short time interval $(t, t + \Delta t]$ given in Table 4.1.

Table 4.1

Transition	Probability of Transition in $(t, t + \Delta]$
$E_1 \to E_1$	$1 - \lambda \Delta t + o(\Delta t)$
$E_1 \to E_0$	$\lambda \Delta t + o(\Delta t)$
$E_0 \to E_1$	$\mu \Delta t + o(\Delta t)$
$E_0 \to E_0$	$1 - \mu \Delta t + o(\Delta t)$

Let $P_1(t)$ = probability that the system is in state E_1 at time t and let $P_0(t)$ = probability that the system is in state E_0 at time t. The difference equations relating $P_0(t)$, $P_1(t)$ with $P_0(t + \Delta t)$, $P_1(t + \Delta.t)$ are

$$P_1(t + \Delta t) = P_1(t)[1 - \lambda \Delta t] + P_0(t)\mu \Delta t + o(\Delta t) \qquad (4.1)$$

$$P_0(t + \Delta t) = P_1(t)\lambda \Delta t + P_0(t)[1 - \mu \Delta t] + o(\Delta t). \qquad (4.2)$$

These equations follow easily from Table 4.1. For example, Equation (4.1) is obtained by noting that the event "system is in E_1 at time $(t + \Delta t)$" occurs in the two mutually exclusive ways:

(i) System is in E_1 at time t and in E_1 at time $t + \Delta t$.

(ii) System is in E_0 at time t and in E_1 at time $t + \Delta t$.

Using Table 4.1, the probability of (i) is $P_1(t)e^{\lambda \Delta t} + o(\Delta t) = P_1(t)[1 - \lambda \Delta t + o(\Delta t)]$. Similarly the probability of (ii) is $P_0(t)(1 - e^{-\mu \Delta t}) + o(\Delta t) = P_0(t)[\mu \Delta t + o(\Delta t)]$. Adding these probabilities gives the right-hand side of Equation (4.1). In a similar way, we can find Equation (4.2). Now, Equation (4.1) is equivalent to

$$\frac{P_1(t + \Delta t) - P_1(t)}{\Delta t} = -\lambda P_1(t) + \mu P_0(t) + \frac{o(\Delta t)}{\Delta t}. \qquad (4.3)$$

As $\Delta t \to 0$, $o(\Delta t)/\Delta t \to 0$ and hence Equation (4.3) becomes

$$P_1'(t) = -\lambda P_1(t) + \mu P_0(t). \qquad (4.4)$$

Similarly Equation (4.2) becomes

$$P_0'(t) = \lambda P_1(t) - \mu P_0(t). \qquad (4.5)$$

Since $P_0(t) + P_1(t) = 1$ for all t, the pair of differential Equations (4.4) and (4.5) are dependent and we need to solve only the first equation, which becomes

$$P_1'(t) = -(\lambda + \mu)P_1(t) + \mu. \qquad (4.6)$$

An explicit solution of the latter requires knowing the state of the system at time $t = 0$. Let us assume that the system starts out in state E_1 (good state) at time $t = 0$, i.e., $P_1(0) = 1$. The solution of Equation (4.6) without the boundary condition is

$$P_1(t) = Ce^{-(\lambda + \mu)t} + \frac{\mu}{\lambda + \mu} \qquad (4.7)$$

for some constant C. Since $P_1(0) = 1$, $C = \lambda/(\lambda + \mu)$, and $P_1(t)$ becomes

$$P_1(t) = \frac{\mu}{\lambda + \mu} + \frac{\lambda}{\lambda + \mu} e^{-(\lambda+\mu)t}, \quad \text{if } P_1(0) = 1. \qquad (4.8)$$

Had we started with the system in state E_0 (bad state), the initial condition would have been $P_1(0) = 0$, and it is easy to verify that in this case the solution to Equation (4.6) with this boundary condition is

$$P_1(t) = \frac{\mu}{\lambda + \mu} - \frac{\mu}{\lambda + \mu} e^{-(\lambda+\mu)t}, \quad \text{if } P_1(0) = 0. \qquad (4.9)$$

Solutions (4.8) and (4.9) are different only because the initial conditions are different. It is awkward to carry along the information about the initial state by the added phrase "if $P_1(0) = 1, 0$" as we have just done. The use of two subscripts, rather than one, eliminates the necessity for doing this. Let $P_{ij}(t)^*$ represent the probability of being in state E_j at time t if the system starts out in state E_i at time $t = 0$. The first subscript now tells us what the initial state was and we can rewrite Equations (4.8) and (4.9) unambiguously as

$$P_{11}(t) = \frac{\mu}{\lambda + \mu} + \frac{\lambda}{\lambda + \mu} e^{-(\lambda+\mu)t} \qquad (4.10)$$

and

$$P_{01}(t) = \frac{\mu}{\lambda + \mu} - \frac{\mu}{\lambda + \mu} e^{-(\lambda+\mu)t}. \qquad (4.11)$$

Since $P_{10}(t) = 1 - P_{11}(t)$ and $P_{00}(t) = 1 - P_{01}(t)$, we have

$$P_{10}(t) = \frac{\lambda}{\lambda + \mu} - \frac{\lambda}{\lambda + \mu} e^{-(\lambda+\mu)t} \qquad (4.12)$$

and

$$P_{00}(t) = \frac{\lambda}{\lambda + \mu} + \frac{\mu}{\lambda + \mu} e^{-(\lambda+\mu)t}. \qquad (4.13)$$

Let us note that

$$p_1 = \lim_{t \to \infty} P_{11}(t) = \lim_{t \to \infty} P_{01}(t) = \frac{\mu}{\lambda + \mu} \qquad (4.14)$$

---

* More generally $P_{ij}(t, t + \tau)$ represents the probability of the system being in state j at time $t + \tau$, given that it was in state E_i at time t. If the process is temporally homogeneous (as it is in the present case), then $P_{ij}(t, t + \tau) = P_{ij}(\tau)$, independently of t.

and

$$p_0 = \lim_{t \to \infty} P_{10}(t) = \lim_{t \to \infty} P_{00}(t) = \frac{\lambda}{\lambda + \mu}.$$

The limiting probabilities p_0 and p_1 are called, respectively, the *steady-state probabilities* of being in E_0 or E_1. These probabilities are independent of the initial state of the system at time $t = 0$.

Equation (4.14) means that the probability of finding a system in state E_1 (E_0) at a time sufficiently far in the future is very close to p_1 (p_0). To see this, consider a system (denoted by ω) chosen at random from an ensemble of possible systems (denoted by Ω) and let the random variable $\xi(t, \omega)$ associated with this system equal 0 (1) if the system is in state E_0 (E_1) at time t. What we have just shown is that $E_\Omega \xi(t, \omega)$, the ensemble mean of $\xi(t, \omega)$, taken over all systems $\omega \in \Omega$ at a fixed time t is given by $P_{11}(t)$ ($P_{01}(t)$) if all systems are initially in E_1 (E_0). An empirical interpretation of this result is as follows: Consider a large number of systems, N, which are initially in state E_1. Let $N_{11}(t)$ ($N_{10}(t)$) be the number of these systems which are in state E_1 (E_0) at time t. Then the empirical frequency $N_{11}(t)/N$ ($N_{10}(t)/N$) will be very close to $P_{11}(t)$ ($P_{10}(t)$) and will converge in probability to $P_{11}(t)$ ($P_{10}(t)$) as $N \to \infty$. To emphasize the fact that $P_{11}(t)$ is really an ensemble average computed at a fixed time point t, we call $P_{11}(t)$ the *point-availability* of a system at time t given that the system was initially in state E_1. As was shown in Equation (4.14) the point-availability (unavailability) $\to p_1$ (p_0) as $t \to \infty$ independently of the initial state of the system.

Another important measure of dependability of a repairable system is *interval-availability*, which is defined as the expected fraction of a given time interval that the system is in a good state. To put this definition in mathematical language, we proceed as follows: For any system ω which is initially in E_1, the associated random function $\xi(t, \omega)$ is a step function with ordinates alternately 1 (when the system is in E_1) and 0 (when the system is in E_0). A particular realization is depicted in Figure 4.1. The total amount of time that the system ω spends in E_1 during the time interval $(a, b]$ is given by the random variable

$$X_{a,b}(\omega) = \int_a^b \xi(t, \omega) dt. \tag{4.15}$$

The random variable $X_{a,b}(\omega)/(b - a)$ is the fraction of the time interval $(a, b]$ that the system ω spends in E_1. The expected fraction of the time interval $(a, b]$ that a system spends in E_1 is by definition the ensemble mean of $X_{a,b}(\omega)/(b - a)$ taken over all systems $\omega \in \Omega$. This ensemble average is denoted as $E_\Omega [X_{a,b}(\omega)] / (b - a)$, where

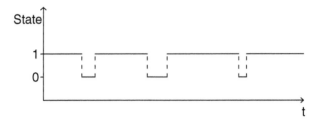

Figure 4.1 *A possible realization over time of a one-unit repairable system, which starts off in a good state.*

$$E_\Omega \left[\frac{X_{a,b}(\omega)}{b-a} \right] = E \left[\frac{1}{b-a} \int_a^b \xi(t,\omega)dt \right] \qquad (4.16)$$

$$= \frac{1}{b-a} \int_a^b E_\Omega \left[\xi(t,\omega) \right] dt = \frac{1}{b-a} \int_a^b P_{11}(t)dt.$$

The key step in Equation (4.16) involves interchanging the order of integration. This can be justified by the fact that $0 \le \xi(t,\omega) \le 1$ is bounded.

Using Equation (4.10) we can obtain an explicit expression for the right-hand side of Equation (4.16). For the particular case where $a = 0, b = T$ the interval availability, which we denote as $(IA_1)_T$ (the subscript 1 means that the system is initially in E_1), is

$$(IA_1)_T = \frac{1}{T} \int_0^T P_{11}(t)dt = \frac{\mu}{\lambda + \mu} + \frac{\lambda}{T(\lambda + \mu)^2} \left[1 - e^{-(\lambda+\mu)T} \right]. \quad (4.17)$$

Similarly if the system is initially in E_0, the interval availability $(IA_0)_T$ is given by

$$(IA_0)_T = \frac{1}{T} \int_0^T P_{01}(t)dt = \frac{\mu}{\lambda + \mu} - \frac{\mu}{T(\lambda + \mu)^2} \left[1 - e^{-(\lambda+\mu)T} \right]. \quad (4.18)$$

Limiting interval availability (also called *long-run system availability* or simply *availability*) is defined as the fraction of time that a system is in a good state over the long run. Availability is usually denoted by the letter A and so we see that for the one unit repairable system

$$A = \lim_{T \to \infty} (IA_1)_T = \lim_{T \to \infty} (IA_0)_T = \frac{\mu}{\lambda + \mu} = p_1. \qquad (4.19)$$

Suppose that we define $R(\tau;t)$, the reliability of a system in the time interval $[t, t+\tau]$ (called *interval reliability* for short), as the probability that the system is in good state throughout the interval $[t, t + \tau]$. Assuming that the system starts off in E_1, it is readily verified that

$$R(\tau;t) = P\{\xi(s,\omega) = 1,\ t \le s \le t+\tau\} \tag{4.20}$$

$$= P\{\xi(s,\omega) = 1\}e^{-\lambda\tau} = P_{11}(t)e^{-\lambda\tau}.$$

Namely, $P_{11}(t)$ is the probability that a randomly chosen system is in E_1 at time t (given that it starts at E_1 at time $t = 0$) and $e^{-\lambda\tau}$ is the (conditional) probability that the system does not fail in the time interval $(t, t+\tau]$ given that the system was in E_1 at time t. Both of these events must occur in order that the system is in state E_1 throughout the time interval $[t, t + \tau]$. Multiplying the probabilities for these two events gives us Equation (4.20). More generally, if the system starts off in state E_0 with probability α_0 and in state E_1 with probability $\alpha_1 = 1 - \alpha_0$, then $\alpha_0 P_{01}(t) + \alpha_1 P_{11}(t)$ is the probability that the system will be in state E_1 at time t and the interval reliability in $[t, t+\tau]$ is

$$R(\tau;t) = (\alpha_0 P_{01}(t) + \alpha_1 P_{11}(t))\, e^{-\lambda\tau}. \tag{4.21}$$

The limiting interval reliability for intervals of length τ is given by

$$R(\tau;\infty) = \lim_{t\to\infty} R(\tau;t) = p_1 e^{-\lambda\tau} = \frac{\mu}{\lambda + \mu}e^{-\lambda\tau}. \tag{4.22}$$

Ergodicity. The fact that limiting point-availability and limiting interval availability are both equal to $p_1 = \mu/(\lambda + \mu)$ is called an *ergodic* property of the stochastic process describing the behavior over time of the one-unit repairable system.

Weak and strong convergence. There are stronger results concerning the stochastic behavior of $X(T,\omega)/T$, the fraction of time spent by a system in a good state in $(0, T]$. We have seen that as $T \to \infty$ the expected value of $X(T,\omega)/T \to p_1$. It can be shown that $X(T,\omega)/T$ converges in probability to p_1, i.e., given any $\epsilon > 0$, however small,

$$\lim_{T\to\infty} P\left\{\left|\frac{X(T,\omega)}{T} - p_1\right| > \epsilon\right\} = 0. \tag{4.23}$$

Equation 4.23 is called *weak convergence* of $X(T,\omega)/T$ to p_1 as $T \to \infty$. *Strong convergence*, namely that $X(T,\omega)/T$ converges to p_1 as $T \to \infty$ for almost all ω, can also be shown. For the reader interested in more details about weak and strong convergence in probability theory, see Billingsley's book [6]. The exact distribution of $X(T, \omega)$ is given by Perry et al. [41].

4.2 Generalizations

The results given up to this point are limited in two respects:

(i) They have been obtained under the assumption that times to failure and times to repair are exponentially distributed.

(ii) They are expectations or averages of one sort or another.

For many reliability applications, neither of these limitations are serious. As a matter of fact, if we are satisfied with long-run averages, then either point-availability or interval availability are equal to MTF/(MTF+MTR) for general time to failure and time to repair distributions. More precisely, let us assume that we have a unit with time to failure distribution $F(x)$ and time to repair distribution $G(y)$. It is assumed that the unit will go into a state of repair as soon as it fails, that repair restores it to its original good state and that it begins operating as soon as repair is completed. Thus one generates an infinite sequence of time intervals (assuming that the unit is originally in a good state) $X_1, Y_1, X_2, Y_2, \ldots, X_n, Y_n, \ldots$. It is assumed that the values $\{X_i\}$ are independently distributed random variables with common c.d.f. $F(x)$ and that the values $\{Y_i\}$ are independently distributed random variables with common c.d.f. $G(x)$. It is then a direct consequence of the law of large numbers that if $E(X) = m_x < \infty$ and $E(Y) = m_y < \infty$, then long-run availability is given by $m_x/(m_x + m_y)$. Empirically this means that if we have observed n times to failure $\{X_1, X_2, \ldots, X_n\}$ and the n associated repair times $\{Y_1, Y_2, \ldots, Y_n\}$ then the ratio $\sum_{i=1}^{n} X_i / (\sum_{i=1}^{n} X_i + \sum_{i=1}^{n} Y_i)$, the observed proportion of time spent by the unit in a good state, will be very close to $m_x/(m_x + m_y)$ for very large n. To find point-availability involves setting up what in Chapter 7 of Cox [9] is called an *alternating renewal process*. For general $F(x)$ and $G(x)$, it is not generally possible to find $P_{ij}(t)$, $i, j = 0, 1$ explicitly, although Laplace-Stieltjes transforms of the $P_{ij}(t)$ are expressible in terms of the Laplace-Stieltjes transforms of $F(x)$ and $G(x)$ (see Section 9.3). However, as $t \to \infty$, it can be shown rigorously that $\lim_{t \to \infty} P_{11}(t) = \lim_{t \to \infty} P_{01}(t) = m_x/(m_x + m_y)$. Thus, just as in the case of the exponential time to failure, exponential time to repair case, point-availability is asymptotically equal to interval availability.

Asymptotic normality. The following central limit theorem due to Takács [50] is useful in practice. Let $X(T)$ be the amount of time that the system is in E_1 and let $Y(T) = T - X(T)$ be the amount of time that the system is in E_0 during the time interval $(0, T]$. Suppose that time to failure is a random variable with mean m_x and variance σ_x^2 and that time to repair is a random variable with mean m_y and variance σ_y^2. Then for large T, $X(T)$ is asymptotically normally distributed, namely,

$$X(T) \sim N \left(\frac{m_x T}{m_x + m_y}, \frac{(m_x^2 \sigma_y^2 + m_y^2 \sigma_x^2) T}{(m_x + m_y)^3} \right).$$

Similarly,

$$Y(T) \sim N \left(\frac{m_y T}{m_x + m_y}, \frac{(m_x^2 \sigma_y^2 + m_y^2 \sigma_x^2) T}{(m_x + m_y)^3} \right).$$

This asymptotic result holds for general failure distribution $F(x)$ and general repair distribution $G(y)$. For the special case of exponential failure and repair distributions, set $\sigma_x^2 = m_x^2$ and $\sigma_y^2 = m_y^2$.

On and off cycles. There are many situations, where a system is not required to perform continuously, but according to some prescribed duty cycle (e.g., on for 8 hours, off for 16 hours) with the pattern repeated indefinitely. Of particular interest are such questions as:

(i) What is the probability that the system will be in a good state at the start of a duty interval?

(ii) What is the probability that the system will be in a good state throughout the duty interval?

(iii) What is the expected interval availability during a duty interval?

For the steady-state answers to these questions, we need to use the method of embedded Markov chains. This is done in Example 7 of Subsection 8.1.7.

4.3 Problems and comments

Problems for Section 4.1

1. Prove that $P_{11}(t) > p_1 = \mu/(\lambda + \mu)$ and that $P_{01}(t) < p_1$ for all $t > 0$.

2. Prove that $(IA_1)_T > p_1$ and $(IA_0)_T < p_1$ for all $T > 0$.

3. If the MTR $= 5$ hours, how large should the MTF be, if we wish the availability to be $\geq .95$?

4. Consider five machines for which the time between adjustments is exponentially distributed with mean $= 30$ minutes and for which the time required to make an adjustment is exponentially distributed with mean $= 3$ minutes. It is assumed that each machine operator adjusts his own machine and that all machines operate independently and are in good operating conditions at 8:00 A.M.

(a) What is the probability that all five machines are operating properly at 2:00 P.M.?

(b) What is the probability that none of the machines are operating properly at 2:00 P.M.?

(c) What is the probability that at least three out of the five machines are operating properly at 2:00 P.M.?

(d) Answer all the above questions with the words "at 2:00 P.M." replaced by "throughout a 3–minute interval beginning at 2:00 P.M.".

5. Consider a one-unit repairable system, with constant failure rate λ and constant repair rate μ. Let us observe the system until we have witnessed n failure time intervals and n associated repair time intervals.
Let $\{X_1, X_2, \ldots, X_n\}$ and $\{Y_1, Y_2, \ldots, Y_n\}$ be the n observed failure and repair time intervals, respectively. Show that

$$\frac{\sum_{i=1}^{n} X_i}{\sum_{i=1}^{n} X_i + \sum_{i=1}^{n} Y_i}$$

tends to $\mu/(\lambda + \mu)$ as $n \to \infty$.

Problems for Section 4.2

1. Suppose that times to failure have mean $m_x = 100$ hours and standard deviation $\sigma_x = 50$ hours and that times to repair have mean $m_y = 1$ hour and standard duration $\sigma_y = 1$ hour.

(a) What is the probability that during a 1,000–hour interval, the total time spent on repair exceeds 15 hours?

(b) What is the probability that during a 1,000–hour interval, the total time spent on repair exceeds 20 hours?

2. Suppose that the MTF $= 200$ hours. Show that an MTR ≤ 5.72 hours is needed, if we want to have a probability of at least 0.9 that the fraction of time that the system is in a good state during $T = 1,000$ hours exceeds $\gamma = 0.95$. Assume the time to failure and time to repair are exponential.

CHAPTER 5

Reliability of a two-unit repairable system

5.1 Steady-state analysis

It may happen that we cannot attain the desired point availability or interval-availability with a system composed of a single repairable unit. It then becomes necessary to improve system performance. This can be done, for example:

(1) By increasing the MTF, decreasing the MTR, or both.
(2) By using additional units as either active or inactive spares.

Which method should be used for improving system performance depends on cost, weight, development time, and other considerations.

In this chapter we consider the second method for improving system performance. We treat in detail the special case of a system composed of two identical units (e.g., two computers), each capable of performing the same function. Successful system performance requires that at least one of these two units be in an operable state and it is considered very undesirable (a system failure) if both units are simultaneously inoperable. If a unit is called inactive, we mean that its failure rate is zero until it is put into service. A unit in service is called active, and its time to failure is assumed to be exponentially distributed with p.d.f. $f(t) = \lambda e^{-\lambda t}$, $t \geq 0$. The time to repair a unit is assumed to be exponentially distributed with p.d.f. $g(t) = \mu e^{-\mu t}$, $t \geq 0$. Failures are assumed to be detectable instantaneously (i.e., there is continuous monitoring) and switching on of an inactive unit so that it becomes active is also assumed to be perfect. We consider four possible cases:

(a) Two units, both active, each having constant failure rate λ and one repairman with constant repair rate μ.

(b) Two units, one unit is active and the other unit is an inactive standby until brought into service. The failure rate of an active unit is λ. The failure rate of an inactive unit is zero until it is brought into service, at which time its failure rate becomes λ. There is one repairman and the repair rate is μ.

(c) Two units, both active, each with failure rate λ; two repairmen, each with repair rate μ (the possibility of two repairmen working on the same failed machine is not allowed).

(d) Two units, one active and one inactive, until called upon when the active unit fails, and two repairmen.[*]

We work out case 2(a) in detail:

Clearly there are three possible system states at any time t: E_2 (both units are good); E_1 (1 unit is good and 1 unit is bad); E_0 (both units are bad).

Assuming exponential failure, exponential repair time distributions means that we have a stochastic process, with (conditional) transition probabilities in the short time interval $(t, t + \Delta t]$ given in Table 5.1.

Table 5.1

Transition	Probability of Transition in $(t, t + \Delta t]$
$E_2 \to E_2$	$1 - 2\lambda\Delta t + o(\Delta t)$
$E_2 \to E_1$	$2\lambda\Delta t + o(\Delta t)$
$E_2 \to E_0$	$o(\Delta t)$
$E_1 \to E_2$	$\mu\Delta t + o(\Delta t)$
$E_1 \to E_1$	$1 - (\lambda + \mu)\Delta t + o(\Delta t)$
$E_1 \to E_0$	$\lambda\Delta t + o(\Delta t)$
$E_0 \to E_2$	$o(\Delta t)$
$E_0 \to E_1$	$\mu\Delta t + o(\Delta t)$
$E_0 \to E_0$	$1 - \mu\Delta t + o(\Delta t)$

These probabilities are easy to derive when we bear in mind that in a short time interval of length Δt a good unit will fail with probability $1 - e^{-\lambda\Delta t} = \lambda\Delta t + o(\Delta t)$, and not fail with probability $e^{-\lambda\Delta t} = 1 - \lambda\Delta t + o(\Delta t)$. A repair on a failed unit will be completed with probability $1 - e^{-\mu\Delta t} = \mu\Delta t + o(\Delta t)$ and not completed with probability $e^{-\mu\Delta t} = 1 - \mu\Delta t + o(\Delta t)$, and two or more repairs or failures (including the possibility of one repair and one failure) will have probability $O((\Delta t)^2) = o(\Delta t)$. Thus, for example, the transition $E_2 \to E_2$ in $(t, t + \Delta t]$ means that both units are in a good state at time t and are both still in a good state Δt time units later. Neglecting events having probability $o(\Delta t)$, this can happen if and only if the event "both units do not

---

[*] In cases 2(b) and 2(d), whenever the system is in state E_2 (both units are good), one of the units is in active service and the other unit is an inactive standby. The two units are assumed to be identical and either unit may be active or inactive.

fail in $(t, t + \Delta t]$" occurs. The probability of this event is equal to $e^{-2\lambda\Delta t} = (1 - \lambda\Delta t + o(\Delta t))^2 = 1 - 2\lambda\Delta t + o(\Delta t)$, the value given in Table 5.1.

Similarly one can verify the other entries in Table 5.1. It is an instructive exercise for the reader to carry out this verification.

Let $P_i(t)$ = probability that the system is in state E_i at time t. The following difference equations express the probability of being in a certain state at time $t + \Delta t$ in terms of the probability of being in various states at time t, ignoring terms of order $o(\Delta t)$.

$$
\begin{aligned}
P_2(t + \Delta t) &= P_2(t)[1 - 2\lambda\Delta t] + P_1(t)\mu\Delta t \\
P_1(t + \Delta t) &= P_2(t)2\lambda\Delta t \quad\quad + P_1(t)[1 - (\lambda + \mu)\Delta t] + P_0(t)\mu\Delta t \\
P_0(t + \Delta t) &= \quad\quad\quad\quad\quad\quad + P_1(t)\lambda\Delta t \quad\quad\quad\quad\quad + P_0(t)[1 - \mu\Delta t].
\end{aligned}
\tag{5.1}
$$

Letting $\Delta t \to 0$, we obtain the differential equations

$$
\begin{aligned}
P_2'(t) &= -2\lambda P_2(t) \quad\quad + \mu P_1(t) \\
P_1'(t) &= \quad 2\lambda P_2(t) - (\lambda + \mu)P_1(t) + \mu P_0(t) \\
P_0'(t) &= \quad\quad\quad\quad\quad\quad \lambda P_1(t) - \mu P_0(t).
\end{aligned}
\tag{5.2}
$$

It can be proved that as $t \to \infty$, $\lim P_i(t) = p_i$, $i = 0, 1, 2$. These limits exist independently of initial conditions, and $p_0 + p_1 + p_2 = 1$. Also $\lim P_i'(t) = 0$. Hence letting $t \to \infty$, the system of differential Equation (5.2) becomes the system of algebraic equations

$$
\begin{aligned}
-2\lambda p_2 \quad\quad + \mu p_1 \quad\quad\quad\quad &= 0 \\
2\lambda p_2 - (\lambda + \mu)p_1 + \mu p_0 &= 0 \\
\lambda p_1 - \mu p_0 &= 0.
\end{aligned}
\tag{5.3}
$$

All three equations are linearly dependent, but any two of them are linearly independent. To completely determine p_0, p_1, and p_2, we add the previously mentioned condition

$$
p_0 + p_1 + p_2 = 1.
\tag{5.4}
$$

From the third row of Equation (5.3), we obtain

$$
p_1 = \frac{\mu}{\lambda}p_0 .
\tag{5.5}
$$

From the first row of Equation (5.3), we obtain

$$
p_2 = \frac{\mu}{2\lambda}p_1 .
\tag{5.6}
$$

Using Equation (5.5), this becomes

$$
p_2 = \frac{1}{2}\left(\frac{\mu}{\lambda}\right)^2 p_0.
\tag{5.7}
$$

Replacing p_1 and p_2 in Equation (5.4) by the right-hand side of Equations (5.5) and (5.7), respectively, we obtain

$$p_0 = \frac{1}{1 + \frac{\mu}{\lambda} + \frac{1}{2}\left(\frac{\mu}{\lambda}\right)^2} . \tag{5.8}$$

If we denote the ratio μ/λ by x, then

$$p_0 = \frac{1}{1 + x + \frac{x^2}{2}} = \frac{2}{(1+x)^2 + 1}, \quad p_1 = x p_0, \quad p_2 = \frac{x^2}{2} p_0 . \tag{5.9}$$

As in the one-unit repairable case, p_i has two probabilistic meanings which can be interpreted empirically as follows:

(i) If we look at a large number N of systems at some fixed time t, then the expected number of systems in state E_i is $NP_i(t)$. As $t \to \infty$, $NP_i(t) \to Np_i$.

(ii) If we look at a system for a long time T, then the expected amount of time spent in E_i is approximately $p_i T$. Specifically, the expected fraction of time the system is in E_i converges to p_i, as $T \to \infty$.

If the system is considered good, when it is in states E_1 or E_2, and bad when it is in state E_0, then using the second interpretation we shall call p_0 the *system unavailability* (the fraction of time that the system spends in the bad state E_0) and

$$1 - p_0 = \frac{(1+x)^2 - 1}{(1+x)^2 + 1} \tag{5.10}$$

the *system availability* A (the fraction of time that the system spends in the good states E_1 or E_2).

It can be verified that for cases (b), (c), and (d), the steady-state probabilities are as follows:

Case (b) Two units, one active, one inactive standby, one repairman. Here, $p_2 = x^2 p_0$, $p_1 = x p_0$ and $p_0 = 1/(1 + x + x^2)$. The latter is the system unavailability. The system availability A is

$$A = 1 - p_0 = \frac{x(1+x)}{1 + x + x^2}. \tag{5.11}$$

Case (c) Two units, both active, two repairmen. Here, $p_2 = x^2 p_0$, $p_1 = 2x p_0$ and $p_0 = 1/(1+x)^2$. The system availability A is

$$A = 1 - p_0 = 1 - \frac{1}{(1+x)^2} = \frac{x(x+2)}{(1+x)^2}. \tag{5.12}$$

Case (d) Two units, one active, one inactive, two repairmen. Here, $p_2 = 2x^2 p_0$, $p_1 = 2x p_0$ and $p_0 = 1/(1 + 2x + 2x^2)$. The system availability A is

$$A = 1 - p_0 = \frac{2x(1 + x)}{(1 + x)^2 + x^2}. \tag{5.13}$$

To give some idea of the improvement in system performance as a result of using two units rather than one, we tabulate the values of the availability and unavailability for the special case $x = \mu/\lambda = 9$.

Table 5.2

System	Availability	Unavailability
One unit, one repairman	9/10	1/10
Two units, both active, one repairman	99/101	2/101
Two units, one active, one inactive		
standby, one repairman	90/91	1/91
Two units, both active, two repairmen	99/100	1/100
Two units, one active, one inactive		
standby, two repairmen	180/181	1/181

5.2 Time-dependent analysis via Laplace transform

In Section 5.1 we obtained the steady-state solution for case 2(a). We now wish to solve the system of differential Equations (5.2) for $P_2(t)$, $P_1(t)$, and $P_0(t)$ subject to the initial conditions $P_2(0) = 1$, $P_1(0) = 0$, $P_0(0) = 0$ (i.e., the system is initially in E_2). There are several known methods to solve these equations and they will be discussed in the following sections. We first show how to use the *Laplace transform methods*. Here we give the necessary facts needed for the problem at hand. To learn more about the Laplace transform see Dyke [17].

5.2.1 Laplace transform method

For any function f, defined on $[0, \infty)$, its *Laplace transform* is denoted by $L\{f\}$ or just f^* and is defined by

$$L\{f\}(s) = f^*(s) := \int_0^\infty e^{-st} f(t) dt,$$

provided the integral is finite. For the reader who is not familiar with the Laplace transform, we give the following basic facts.

(i) Uniqueness: $f_1(t) = f_2(t)$ for almost all $t \geq 0$ $\Leftrightarrow$ $f_1^* \equiv f_2^*$.

(ii) Linearity: $(\alpha f_1 + \beta f_2)^* = \alpha f_1^* + \beta f_2^*$ for any two constants α, β.

(iii) If f' is the derivative of f, then

$$f'^*(s) = -f(0) + sf^*(s).$$

(iv) For constants c, α, λ $(\alpha, \lambda \geq 0)$,

$$f(t) = ct^{\alpha}e^{-\lambda t} \ (t \geq 0) \quad \Leftrightarrow \quad f^*(s) = \frac{c\Gamma(\alpha+1)}{(s+\lambda)^{\alpha+1}} \ (s > -\lambda).$$

It is worthwhile to note the special cases of (iv)

$$f(t) \equiv c \quad \Leftrightarrow \quad f^*(s) = \frac{c}{s}$$

and

$$f(t) = e^{-\lambda t} \quad \Leftrightarrow \quad f^*(s) = \frac{1}{s+\lambda}.$$

Using the notation we have just defined, $P_i^*(s)$ is the Laplace transform of $P_i(t)$ and it follows that

$$L\{P_i'\}(s) = sP_i^*(s) - P_i(0). \tag{5.14}$$

In particular, if $P_2(0) = 1$, $P_1(0) = 0$, $P_0(0) = 0$, we have

$$\begin{aligned} L\{P_2'\}(s) &= sP_2^*(s) - 1 \\ L\{P_1'\}(s) &= sP_1^*(s) \\ L\{P_0'\}(s) &= sP_0^*(s). \end{aligned} \tag{5.15}$$

Applying the Laplace transform to both sides of the set of differential Equation (5.2), we obtain the set of linear equations

$$\begin{aligned} (s+2\lambda)P_2^*(s) \qquad -\mu P_1^*(s) \qquad\qquad &= 1 \\ -2\lambda P_2^*(s) +(s+\lambda+\mu)P_1^*(s) \qquad -\mu P_0^*(s) &= 0 \\ -\lambda P_1^*(s) +(s+\mu)P_0^*(s) &= 0. \end{aligned} \tag{5.16}$$

Solving for $P_2^*(s)$, $P_1^*(s)$, $P_0^*(s)$, we obtain

$$P_2^*(s) = \frac{1}{\Delta(s)} \begin{vmatrix} 1 & -\mu & 0 \\ 0 & s+\lambda+\mu & -\mu \\ 0 & -\lambda & s+\mu \end{vmatrix}$$

$$P_1^*(s) = \frac{1}{\Delta(s)} \begin{vmatrix} s+2\lambda & 1 & 0 \\ -2\lambda & 0 & -\mu \\ 0 & 0 & s+\mu \end{vmatrix} \tag{5.17}$$

$$P_0^*(s) = \frac{1}{\Delta(s)} \begin{vmatrix} s+2\lambda & -\mu & 1 \\ -2\lambda & s+\lambda+\mu & 0 \\ 0 & -\lambda & 0 \end{vmatrix},$$

where

$$\Delta(s) = \begin{vmatrix} s + 2\lambda & -\mu & 0 \\ -2\lambda & s + \lambda + \mu & -\mu \\ 0 & -\lambda & s + \mu \end{vmatrix}$$

$$= s^3 + (3\lambda + 2\mu)s^2 + (2\lambda^2 + 2\lambda\mu + \mu^2)s$$

$$= s(s - s_1)(s - s_2)$$

is the *characteristic polynomial* and where

$$s_1 = \frac{-(3\lambda + 2\mu) + \sqrt{\lambda^2 + 4\lambda\mu}}{2} \tag{5.18}$$

and

$$s_2 = \frac{-(3\lambda + 2\mu) - \sqrt{\lambda^2 + 4\lambda\mu}}{2}$$

are the *characteristic roots* (also called *eigen values*). It is obvious that $s_1 < 0$ and $s_2 < 0$. Substituting Equation (5.18) into Equation (5.17), we have

$$P_2^*(s) = \frac{s^2 + (\lambda + 2\mu)s + \mu^2}{s(s - s_1)(s - s_2)},$$

$$P_1^*(s) = \frac{2\lambda(s + \mu)}{s(s - s_1)(s - s_2)} \tag{5.19}$$

and

$$P_0^*(s) = \frac{2\lambda^2}{s(s - s_1)(s - s_2)}.$$

The next step is to expand each ratio in Equation (5.19) into partial fractions, that is, we write $P_2^*(s)$ as a sum

$$P_2^*(s) = \frac{\alpha_2}{s} + \frac{\beta_2}{s - s_1} + \frac{\gamma_2}{s - s_2}. \tag{5.20}$$

The coefficients $\alpha_2, \beta_2, \gamma_2$ are extracted from the identity of the numerator of $P_2^*(s)$ in Equation (5.19) and the numerator of Equation (5.20) when all three fractions are combined on a common denominator. In our case the identity is

$$\alpha_2(s - s_1)(s - s_2) + \beta_2 s(s - s_2) + \gamma_2 s(s - s_1) = s^2 + (\lambda + 2\mu)s + \mu^2. \tag{5.21}$$

An alternative way is to compute these coefficients via

$$\alpha_2 = sP_2^*(s)|_{s=0}, \quad \beta_2 = (s - s_1)P_2^*(s)|_{s=s_1}, \quad \gamma_2 = (s - s_2)P_2^*(s)|_{s=s_2}. \tag{5.22}$$

Using either Equation (5.21) or (5.22), we get

$$\alpha_2 = \frac{\mu^2}{s_1 s_2} = \frac{\mu^2}{2\lambda^2 + 2\lambda\mu + \mu^2} = p_2,$$

$$\beta_2 = \frac{(s_1 + \mu)^2 + \lambda s_1}{s_1(s_1 - s_2)}$$

and

$$\gamma_2 = \frac{(s_2 + \mu)^2 + \lambda s_2}{s_2(s_2 - s_1)}.$$

Similarly, repeating the process for $P_1^*(s)$ and $P_0^*(s)$ (though the latter is not necessary since we can solve the final equation for $P_0(t)$ from $P_2 + P_1 + P_0 = 1$), we get

$$P_1^*(s) = \frac{\alpha_1}{s} + \frac{\beta_1}{s - s_1} + \frac{\gamma_1}{s - s_2},$$

where

$$\alpha_1 = \frac{2\lambda\mu}{2\lambda^2 + 2\lambda\mu + \mu^2} = p_1, \quad \beta_1 = \frac{2\lambda(s_1 + \lambda)}{s_1(s_1 - s_2)}, \quad \gamma_1 = \frac{2\lambda(s_2 + \lambda)}{s_2(s_2 - s_1)}$$

and finally,

$$P_0^*(s) = \frac{\alpha_0}{s} + \frac{\beta_0}{s - s_1} + \frac{\gamma_0}{s - s_2},$$

where

$$\alpha_0 = \frac{2\lambda^2}{2\lambda^2 + 2\lambda\mu + \mu^2} = p_0, \quad \beta_0 = \frac{2\lambda^2}{s_1(s_1 - s_2)}, \quad \gamma_0 = \frac{2\lambda^2}{s_2(s_2 - s_1)}.$$

To find the $P_i(t)$ from the $P_i^*(s)$, we recall that

$$L\{1\}(s) = \int_0^\infty e^{-st} dt = \frac{1}{s} \qquad (5.23)$$

and

$$L\{e^{s_i t}\}(s) = \int_0^\infty e^{-st} e^{s_i t} dt = \int_0^\infty e^{-(s - s_i)t} dt = \frac{1}{s - s_i}.$$

Consequently, it follows that the time dependent solutions for $P_2(t)$, $P_1(t)$, $P_0(t)$ are:

$$\begin{aligned} P_2(t) &= p_2 + \beta_2 e^{s_1 t} + \gamma_2 e^{s_2 t} \\ P_1(t) &= p_1 + \beta_1 e^{s_1 t} + \gamma_1 e^{s_2 t} \\ P_0(t) &= p_0 + \beta_0 e^{s_1 t} + \gamma_0 e^{s_2 t}. \end{aligned} \qquad (5.24)$$

It should be noted that $P_0(t)$ is the probability the system is in state 0 at time t, given it started at state 2 at time 0. If T is the time when the system enters state 0 for the first time, and we ask what is $P\{T \le t\}$, then $P_0(t)$ is not the answer. This matter is discussed in Chapter 7. Since s_1 and s_2 are both negative, $\lim_{t \to \infty} P_i(t) = p_i$, the steady-state solution. The transient part of the solution damps out exponentially.

To make explicit that $P_2(t)$, $P_1(t)$, and $P_0(t)$ are, respectively, the probabilities of being in states E_2, E_1 and E_0 given that the system was initially in state E_2, we write $P_2(t)$ as $P_{22}(t)$, $P_1(t)$ as $P_{21}(t)$, and $P_0(t)$ as $P_{20}(t)$. More generally $P_{ij}(t)$ represents the probability of being in state E_j at time t given that the initial state at time $t = 0$ was E_i. Thus $P_{12}(t)$, $P_{11}(t)$, and $P_{10}(t)$ are, respectively, the probabilities of being in states E_2, E_1 and E_0 at time t, given that the initial state was E_1. Similarly $P_{02}(t)$, $P_{01}(t)$, and $P_{00}(t)$ are, respectively, the probabilities of being in states E_2, E_1 and E_0 at time t, given that the initial state was E_0.

If we denote the Laplace transform of P_{ij} by

$$P_{ij}^*(s) = L\{P_{ij}\}(s) = \int_0^\infty e^{-st} P_{ij}(t) dt, \tag{5.25}$$

then Equation (5.16) can be rewritten as

$$
\begin{array}{lll}
(s + 2\lambda) P_{22}^*(s) & -\mu P_{21}^*(s) & = 1 \\
-2\lambda P_{22}^*(s) + (s + \lambda + \mu) P_{21}^*(s) & -\mu P_{20}^*(s) = 0 \\
& -\lambda P_{21}^*(s) + (s + \mu) P_{20}^*(s) = 0.
\end{array} \tag{5.26}
$$

Similarly, if we wish to solve the set of differential Equation (5.2) subject to the initial conditions that $P_2(0) = 0$, $P_1(0) = 1$, $P_0(0) = 0$ (corresponding to initial state E_1), we obtain the set of equations

$$
\begin{array}{lll}
(s + 2\lambda) P_{12}^*(s) & -\mu P_{11}^*(s) & = 0 \\
-2\lambda P_{12}^*(s) + (s + \lambda + \mu) P_{11}^*(s) & -\mu P_{10}^*(s) = 1 \\
& -\lambda P_{11}^*(s) + (s + \mu) P_{10}^*(s) = 0.
\end{array} \tag{5.27}
$$

If the initial conditions are $P_2(0) = 0$, $P_1(0) = 0$, $P_0(0) = 1$ (corresponding to initial state E_0), we obtain the set of equations

$$
\begin{array}{lll}
(s + 2\lambda) P_{02}^*(s) & -\mu P_{01}^*(s) & = 0 \\
-2\lambda P_{02}^*(s) + (s + \lambda + \mu) P_{01}^*(s) & -\mu P_{00}^*(s) = 0 \\
& -\lambda P_{01}^*(s) + (s + \mu) P_{00}^*(s) = 1.
\end{array} \tag{5.28}
$$

The introduction of matrix notation makes it possible to write the system of differential Equation (5.2) and the systems of Equations (5.26), (5.27), and (5.28) very concisely. Let the matrix Q be defined as the array

$$
Q = \begin{pmatrix}
-\mu & \mu & 0 \\
\lambda & -(\lambda + \mu) & \mu \\
0 & 2\lambda & -2\lambda
\end{pmatrix}. \tag{5.29}
$$

This matrix is known as the *infinitesimal generator* of the process. Let the matrix $\mathbb{P}(t)$ be defined as

$$\mathbb{P}(t) = \begin{pmatrix} P_{00}(t) & P_{01}(t) & P_{02}(t) \\ P_{10}(t) & P_{11}(t) & P_{12}(t) \\ P_{20}(t) & P_{21}(t) & P_{22}(t) \end{pmatrix}, \tag{5.30}$$

where the first, second, and third rows of $\mathbb{P}(t)$ are vectors corresponding to the solution of Equation (5.2) when the initial states are E_0, E_1 and E_2, respectively. This is the *transition probabilities matrix* of the process. Similarly the matrix $\mathbb{P}'(t)$ is obtained from $\mathbb{P}(t)$ by replacing each $P_{ij}(t)$ by its derivative $P'_{ij}(t)$. Hence, Equation (5.2) can be written as

$$\mathbb{P}'(t) = \mathbb{P}(t)Q. \tag{5.31}$$

If we define the identity matrix I as

$$I = \begin{pmatrix} 1 & 0 & 0 \\ 0 & 1 & 0 \\ 0 & 0 & 1 \end{pmatrix} \tag{5.32}$$

and the matrix of Laplace transforms $P_{ij}^*(s)$ as

$$\mathbb{P}^*(s) = \begin{pmatrix} P_{00}^*(s) & P_{01}^*(s) & P_{02}^*(s) \\ P_{10}^*(s) & P_{11}^*(s) & P_{12}^*(s) \\ P_{20}^*(s) & P_{21}^*(s) & P_{22}^*(s) \end{pmatrix} \tag{5.33}$$

then, the systems of Equations (5.26), (5.27), and (5.28) can be written as

$$\mathbb{P}^*(s)(sI - Q) := \mathbb{P}^*(s)\mathbb{B}(s) = I, \tag{5.34}$$

where $\mathbb{B}(s) = sI - Q$. Therefore

$$\mathbb{P}^*(s) = \mathbb{B}^{-1}(s). \tag{5.35}$$

Let $b_{ij}(s)$ be the ij element of $\mathbb{B}(s)$. Then, it follows from Equation (5.35) that $P_{ij}^*(s)$, the Laplace transform of $P_{ij}(t)$, satisfies the equation

$$P_{ij}^*(s) = \frac{B_{ji}(s)}{\Delta(s)}, \tag{5.36}$$

where $B_{ji}(s)$ is the cofactor of $b_{ji}(s)$ and † where $\Delta(s) = |\mathbb{B}(s)|$ is the determinant of the matrix $\mathbb{B}(s)$. The cofactor $B_{ji}(s)$ is a polynomial in s of degree less than or equal to two and $\Delta(s)$ is a polynomial in s of degree three. It is readily verified that one of the roots of $\Delta(s)$ is $s = 0$, since

---

† The cofactor of $b_{ji}(s)$ is the determinant (with the proper sign) of the matrix obtained by erasing the jth row and the ith column of $\mathbb{B}(s)$.

$\Delta(0) = |-Q| = 0$, and it can further be shown that the other two roots s_1 and s_2 must be negative. Thus, the ratio $P_{ij}^*(s) = B_{ji}(s)/\Delta(s)$ can be expanded in partial fractions, as we did in Equation (5.20), and expressed in the form $\alpha_{ij}/s + \beta_{ij}/(s - s_1) + \gamma_{ij}/(s - s_2)$. Inverting the Laplace transform of $P_{ij}^*(s)$ yields

$$\begin{aligned} P_{ij}(t) &= \alpha_{ij} + \beta_{ij}e^{s_1t} + \gamma_{ij}e^{s_2t} \\ &= p_j + \beta_{ij}e^{s_1t} + \gamma_{ij}e^{s_2t}, \end{aligned} \quad (5.37)$$

since $\lim_{t\to\infty} P_{ij}(t) = p_j$ independently of i. This can be conveniently written in matrix notation. For each initial state E_i, $i = 0, 1, 2$, let $\mathbb{C}_i$ be the 3×3 matrix of coefficients

$$\mathbb{C}_i = \begin{pmatrix} p_0 & p_1 & p_2 \\ \beta_{i0} & \beta_{i1} & \beta_{i2} \\ \gamma_{i0} & \gamma_{i1} & \gamma_{i2} \end{pmatrix}$$

and let $\mathbf{y}(t) = (1, e^{s_1t}, e^{s_2t})$. Then the row vector of probabilities $\mathbf{P}_i(t) = (P_{i0}(t), P_{i1}(t), P_{i2}(t))$ (the ith row of the matrix $\mathbb{P}(t)$) is given by

$$\mathbf{P}_i(t) = \mathbf{y}(t)\mathbb{C}_i. \quad (5.38)$$

It can be proved that for every j, independently of i,

$$p_j = \frac{Q_{ji}}{\sum_k Q_{ki}}, \quad (5.39)$$

where Q_{ji} is the cofactor of q_{ji}, the ji element of the matrix Q.

5.2.2 A numerical example

Consider model 2(a), two identical active units, each one with failure rate $\lambda = 1$ and one repairman with repair rate $\mu = 6$. Then, the matrix Q is equal to

$$Q = \begin{pmatrix} -6 & 6 & 0 \\ 1 & -7 & 6 \\ 0 & 2 & -2 \end{pmatrix},$$

the matrix $\mathbb{B}(s)$ is equal to

$$\mathbb{B}(s) = \begin{pmatrix} s+6 & -6 & 0 \\ -1 & s+7 & -6 \\ 0 & -2 & s+2 \end{pmatrix}$$

and its determinant is given by

$$\Delta(s) = |\mathbb{B}(s)| = s(s+5)(s+10) = s^3 + 15s^2 + 50s.$$

The next step is to compute all the cofactors $B_{ij}(s)$ of the matrix $\mathbb{B}(s)$. Each one is a polynomial of degree 2 or less. Let $\mathcal{B}(s)$ be the matrix whose ij element is $B_{ij}(s)$. Then, careful calculation yields the matrix

$$
\mathcal{B}(s) = \begin{pmatrix}
s^2 + 9s + 2 & s + 2 & 2 \\
6s + 12 & s^2 + 8s + 12 & 2s + 12 \\
36 & 6s + 36 & s^2 + 13s + 36
\end{pmatrix}.
$$

By Equation (5.36) we have

$$
\mathbb{P}^*(s) = \frac{1}{\Delta(s)} \mathcal{B}^T(s),
$$

where T stands for *transpose*.

The next step is to calculate the entries of the matrices $\mathbb{C}_i$, $i = 0, 1, 2$. That is, each ratio $B_{ji}(s)/\Delta(s)$ is expanded in partial fractions $p_j/s + \beta_{ij}/(s + 5) + \gamma_{ij}/(s + 10)$, and the coefficients are determined as in Equation (5.22). Doing this for all i, j, we obtain the matrices $\mathbb{C}_i$

$$
\mathbb{C}_0 = \frac{1}{25} \begin{pmatrix}
1 & 6 & 18 \\
18 & 18 & -36 \\
6 & -24 & 18
\end{pmatrix}
$$

$$
\mathbb{C}_1 = \frac{1}{25} \begin{pmatrix}
1 & 6 & 18 \\
3 & 3 & -6 \\
-4 & 16 & -12
\end{pmatrix}
$$

and

$$
\mathbb{C}_2 = \frac{1}{25} \begin{pmatrix}
1 & 6 & 18 \\
-2 & -2 & 4 \\
1 & -4 & 3
\end{pmatrix}.
$$

Calculating the row vectors $\mathbf{P}_i(t) = \mathbf{y}(t)\mathbb{C}_i = (1, e^{-5t}, e^{-10t})\mathbb{C}_i$ for $i = 0, 1, 2$, and writing them one under the other, gives the matrix $\mathbb{P}(t)$. Due to space limitations, we give the matrix $25\mathbb{P}(t)$ below:

$$
\begin{pmatrix}
1 + 18e^{-5t} + 6e^{-10t} & 6 + 18e^{-5t} - 24e^{-10t} & 18 - 36e^{-5t} + 18e^{-10t} \\
1 + 3e^{-5t} - 4e^{-10t} & 6 + 3e^{-5t} + 16e^{-10t} & 18 - 6e^{-5t} - 12e^{-10t} \\
1 - 2e^{-5t} + e^{-10t} & 6 - 2e^{-5t} - 4e^{-10t} & 18 + 4e^{-5t} + 3e^{-10t}
\end{pmatrix}.
$$

To see the validity of Equation (5.39), we observe that

$$
Q_{00} = Q_{01} = Q_{02} = 2, \quad Q_{10} = Q_{11} = Q_{12} = 12, \quad Q_{20} = Q_{21} = Q_{22} = 36
$$

hence,

$$
p_0 = \frac{2}{2 + 12 + 36}, \quad p_1 = \frac{12}{2 + 12 + 36}, \quad p_2 = \frac{36}{2 + 12 + 36},
$$

resulting in agreement with the time dependent result. Finally, the system availability is

$$1 - p_0 = 1 - \frac{1}{25} = \frac{24}{25} = \frac{(1+6)^2 - 1}{(1+6)^2 + 1},$$

where the last term is from Equation (5.10).

5.3 On Model 2(c)

It should be noted that for the case of two machines and two repairmen, the
time dependent solutions for $P_{ij}(t)$ can be expressed very easily in terms of
$p_{00}(t)$, $p_{01}(t)$, $p_{10}(t)$ and $p_{11}(t)$ of the one-unit repairable system.[‡] The reason
for this is that the two machines and two repairmen can be paired off as two
independent machine–repairman combinations, i.e., as two independent one-
unit repairable systems. Clearly the two-unit system can be in:

E_2, if and only if each unit is in a good state.

E_1, if and only if one of the units is in a good state and the other is in a bad
state.

E_0, if and only if each unit is in a bad state.

Consequently, the $P_{ij}(t)$, $i, j = 0, 1, 2$ are

$$P_{22}(t) = p_{11}^2(t), \qquad P_{21}(t) = 2p_{11}(t)p_{10}(t),$$
$$P_{20}(t) = p_{10}^2(t).$$

$$P_{12}(t) = p_{11}(t)p_{01}(t), \quad P_{11}(t) = p_{10}(t)p_{01}(t) + p_{11}(t)p_{00}(t),$$
$$P_{10}(t) = p_{10}(t)p_{00}(t). \tag{5.40}$$

$$P_{02}(t) = p_{01}^2(t), \qquad P_{01}(t) = 2p_{01}(t)p_{00}(t),$$
$$P_{00}(t) = p_{00}^2(t).$$

Note, $\Sigma_j P_{ij}(t) = 1$ for all i. These formulae for $P_{ij}(t)$ are valid for general
unit time to failure and unit time to repair distributions. For the special case
of exponentially distributed unit times to failure and times to repair the $p_{ij}(t)$
are given by formulae (4.8)–(4.13). The $P_{ij}(t)$ for general time to failure and
repair distributions are given in Section 9.3.

[‡] Note that to avoid confusion we have used lowercase letters $p_{ij}(t)$, $i, j = 0, 1$, for the one-unit
repairable system.

If the mean time to failure for a unit is $m_x < \infty$ and the mean time to repair for a unit is $m_y < \infty$, it has been shown in Chapter 4 that

$$\lim_{t \to \infty} p_{11}(t) = \lim_{t \to \infty} p_{01}(t) = \frac{m_x}{m_x + m_y}$$

and

$$\lim_{t \to \infty} p_{10}(t) = \lim_{t \to \infty} p_{00}(t) = \frac{m_y}{m_x + m_y}.$$

Consequently, steady-state probabilities for Model 2(c) are given by

$$p_2 = \lim_{t \to \infty} P_{22}(t) = \lim_{t \to \infty} P_{12}(t) = \lim_{t \to \infty} P_{02}(t) = \left(\frac{m_x}{m_x + m_y}\right)^2$$

$$p_1 = \lim_{t \to \infty} P_{21}(t) = \lim_{t \to \infty} P_{11}(t) = \lim_{t \to \infty} P_{01}(t) = \frac{2m_x m_y}{(m_x + m_y)^2}$$

and

$$p_0 = \lim_{t \to \infty} P_{20}(t) = \lim_{t \to \infty} P_{10}(t) = \lim_{t \to \infty} P_{00}(t) = \left(\frac{m_y}{m_x + m_y}\right)^2.$$

In particular, if E_0 is considered a failed state for the system, the long run system unavailability $= p_0 = [m_y/(m_x + m_y)]^2$. For the special case of exponential time to failure, exponential time to repair, $m_x = 1/\lambda$, $m_y = 1/\mu$ and $p_0 = (\lambda/(\lambda + \mu))^2$.

In Section 7.3 we further discuss the two-unit repairable system, where the time to repair has a general distribution.

5.4 Problems and Comments

Problems for Section 5.1

1. Find p_0, p_1, and p_2 for the case of two active units and two repairmen (Model 2(c)) using the result that for a one-unit repairable system the availability is $x/(1 + x)$ and the unavailability $1/(1 + x)$.
   Hint: Use the binomial distribution.

2. Find the availability and unavailability for a two-unit system with states E_2 (two good units, one active with failure rate λ_a and one standby with failure rate $0 \le \lambda_b \le \lambda_a$); E_1 (one good unit and with associated failure rate λ_a and one bad unit); E_0 (two bad units). States E_1 or E_2 are considered good states for the system.

   (a) Solve if there is one repairman with repair rate μ.
   (b) Solve if there are two repairmen, each having repair rate μ.

3. Table 5.1 gives the transition probabilities of going from one state to another in the time interval $(t, t+\Delta t]$ for Case 2(a). Write out similar tables for cases 2(b), 2(c), and 2(d).

4. Compute $x = \mu/\lambda$ so that the expected total time that both units are down over 10, 000 hours of operation is less than or equal to 4 hours. Do this for cases 2(a), 2(b), 2(c), and 2(d).

Problems for Section 5.2

1. Use Laplace transform methods to find $P_{00}(t)$, $P_{01}(t)$, $P_{10}(t)$ and $P_{11}(t)$ for the one-unit repairable system discussed in Chapter 4.

2. Use Laplace transform methods to find $P_n(t) = P\left(N(t) = n\right)$ for a homogeneous Poisson process with parameter λ.
   Hint: $L\left\{t^n e^{-\lambda t}\right\}(s) = n!/(s + \lambda)^{n+1}$.

3. Find $P_{ij}^*(s)$, the Laplace transform of $P_{ij}(t)$, for case 2(b) $-$ 2 units, one active with constant failure rate λ and one inactive ($\lambda = 0$) standby, and one repairman with constant repair rate μ.

4. Find α_{ij}, β_{ij} and γ_{ij} in Equation (5.37).

5. Derive Formula (5.39).

6. Verify that in Equations (5.19) $\lim_{s \to 0} sP_i^*(s) = \lim_{t \to \infty} P_i(t)$.

7. Compute the matrix $\mathbb{P}(t)$ for case 2(b) with $\lambda = 1$ and $\mu = 9$.

CHAPTER 6

Continuous-time Markov chains

6.1 The general case

6.1.1 Definition and notation

The one-unit and two-unit repairable systems are special cases of *continuous-time Markov chains*, also known as *Markov jump processes*. The full theory is discussed in Karlin and Taylor [35] and Stroock [48]. Here we present the essential facts, relevant to our systems reliability.

Consider a system with possible states $E_i \in S$ (finite or infinite). In many cases $E_i = i$, as in n-unit repairable system, where i is the number of operational units or in queuing models, where i is the number of customers in the system. But this is not always the case. For instance, think of a two-unit repairable system with one repairman. Suppose unit 1 is exposed to a failure rate λ_1 and unit 2 to a failure rate λ_2. Here, state 2 is indeed telling us how many units are operational. But if one unit has failed and is now under repair, saying that the system is in state 1 does not give the full information. One has to distinguish between "unit 1 is under repair and unit 2 is operational" (state E_1', say) and "unit 2 is under repair and unit 1 is operational" (state E_1''). Similarly, E_0' is the state in which both units have failed but unit 1 is under repair and E_0'' is the state in which both units have failed but unit 2 is under repair. In this system we have 5 different states. For simplicity we will say that the system is in state i when we really mean state E_i. The state of the system at time t is denoted by $\xi(t) \in S$. The stochastic process $\{\xi(t) : t \geq 0\}$ is said to be a continuous-time Markov chain if for all $i, j \in S$,

$$P\{\xi(t+s) = j \mid \xi(s) = i, \, \xi(u), \, 0 \leq u < s\}$$

$$= P\{\xi(t+s) = j \mid \xi(s) = i\}.$$

(6.1)

This is the *Markov property*, namely, given the present state (at time s) and all the past history of the system before time s, the distribution of the future state (at time $s + t$) depends only on the present state. If we further assume that

Equation (6.1) does not depend on s, then the system or the process is said to have *stationary* (or *homogeneous*) *transition probabilities*. Indeed, this is what we assume throughout and thus the transition probability in Equation (6.1) is denoted by $P_{ij}(t)$.

Suppose that our system starts out in state i at time $t = 0$ and we know that until time s it never left the initial state. What is the probability it will stay there t more time-units? Let T_i be the (random) time the system stays in state i until it makes a transition to another state. Then, by the stationarity assumption,

$$P\{T_i > s + t \mid T_i > s\} = P\{T_i > t\} \quad (s, t \geq 0). \tag{6.2}$$

As is well known, the only nontrivial (continuous) distribution which satisfies the memoryless condition (6.2) is the exponential distribution with, say, rate q_i $(0 < q_i < \infty)$. Hence, the process can be described as follows:

Whenever the system enters into state i, it stays there an exponential amount of time with parameter q_i. It then makes a transition into state j with probability p_{ij} $(\Sigma_{j \neq i} \, p_{ij} = 1)$. The system then stays there an exponential amount of time with parameter q_j, etc.

So, we see that the assumption in Chapters 4 and 5 that time to failure and time to repair are exponential is not just for convenience, but rather quite natural. Let $\lambda_{ij} = q_i p_{ij}$, then λ_{ij} is the rate at which the system makes a transition from state i to state j. Given that the system is in state i, we can think of the transition process as a result of independent competing Poisson processes, one for each j $(j \neq i)$ with constant rate λ_{ij} and time to first "event" T_{ij}. The time the system remains in state i is then

$$T_i = \min_{j \neq i} T_{ij} \, ,$$

which is exponential with rate $\Sigma_{j \neq i} \lambda_{ij} = q_i$. It follows that the λ_{ij} completely determines the transition probabilities of the Markov chain. The system is often described graphically by a *transitions and rates diagram* as in Figure 6.1. Here we see an example of a system with 4 states. If $\lambda_{ij} > 0$ for a pair i, j, then we say that state j is *directly accessible* from state i and we draw an arrow from state i to state j. In the present example, every two states are accessible both ways, but not necessarily in one transition. Such a chain is called *irreducible*. If $q_i = 0$ for some state i (not in this example), it means that once the system enters this state, it will never leave it. Such a state is called *absorbing*.

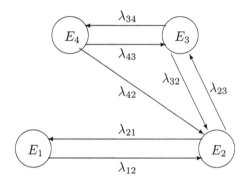

Figure 6.1 *Transitions and rates diagram of a system with 4 states.*

6.1.2 The transition probabilities

The Markov property and the stationarity assumption imply that for all $i, j \in \mathcal{S}$, $s, t \geq 0$,

$$P_{ij}(t + s) = \sum_{k \in \mathcal{S}} P_{ik}(t) P_{kj}(s), \tag{6.3}$$

known as the *Chapman-Kolmogorov equations.* Suppose there are exactly n possible states. Then, in matrix notation this is

$$\mathbb{P}(t + s) = \mathbb{P}(t)\mathbb{P}(s). \tag{6.4}$$

As in the one-dimensional case, the solution to Equation (6.4) is

$$\mathbb{P}(t) = e^{Qt} = I + \sum_{k=1}^{\infty} (Qt)^k / k!, \tag{6.5}$$

where Q is some $n \times n$ matrix and I is the $n \times n$ identity matrix. Differentiating with respect to t, one gets the differential equations

$$\mathbb{P}'(t) = \sum_{k=0}^{\infty} (Qt)^k / k! Q = Q \sum_{k=0}^{\infty} (Qt)^k / k! = \mathbb{P}(t)Q = Q\mathbb{P}(t). \tag{6.6}$$

The equalities $\mathbb{P}'(t) = \mathbb{P}(t)Q$ and $\mathbb{P}'(t) = Q\mathbb{P}(t)$ are called the *Kolmogorov forward and backward equations,* respectively. Since $\mathbb{P}(0) = I$, we conclude that

$$Q = \mathbb{P}'(0). \tag{6.7}$$

To evaluate the entries q_{ij} of the matrix Q, we use Equation (6.3), where we put $s = \Delta \downarrow 0$. The result is

$$P_{ij}(t + \Delta) = \sum_{k \neq j} P_{ik}(t) \lambda_{kj} \Delta + P_{ij}(t)(1 - q_j \Delta) + o(\Delta),$$

which leads to

$$P'_{ij}(t) = \sum_{k \neq j} P_{ik}(t)\lambda_{kj} - q_j P_{ij}(t). \tag{6.8}$$

Since $\mathbb{P}(0) = I$ and $\mathbb{P}'(0) = Q$, we get the entries of Q as

$$q_{ij} = \lambda_{ij} \ (i \neq j), \quad q_{ii} = -q_i = -\Sigma_{j \neq i}\lambda_{ij}. \tag{6.9}$$

Hence, we see again that the rates λ_{ij} determine the transition probabilities. The matrix Q is known as the *infinitesimal generator* of the Markov chain, also called the *transition rates matrix*.

Finally, to evaluate $P\{\xi(t) = j\}$, one has to assume an initial distribution on the state of the system at time 0. Denote it by the row-vector $\boldsymbol{\pi}$, whose j^{th} component is $\pi_j = P\{\xi(0) = j\}$. Then,

$$P_j(t) := P\{\xi(t) = j\} = \sum_{i \in \mathcal{S}} \pi_i P_{ij}(t)$$

and if $\mathbf{P}(t) = (P_1(t), P_2(t), \dots)$ is the row vector of the state distribution at time t, then

$$\mathbf{P}(t) = \boldsymbol{\pi}\mathbb{P}(t). \tag{6.10}$$

6.1.3 Computation of the matrix $\mathbb{P}(t)$

Given a continuous-time Markov chain, it is practically a good idea to draw the states and transition rates diagram and determine the λ_{ij}, which in turn, determine the transition rates matrix Q. It is assumed that there are exactly n states, indexed by $0, 1, \dots, n - 1$. We present here three methods to compute the transition probabilities matrix $\mathbb{P}(t)$.

The Laplace transform method. This method has been demonstrated in Subsection 5.2.1 for 3-state systems. Here is a list of steps required by the method.

Step 1. Compute the characteristic polynomial of Q

$$\Delta(s) = |sI - Q| = |\mathbb{B}(s)|$$

and find its roots $s_0 = 0, s_1, \dots, s_{n-1}$.

Step 2. Compute all cofactors $B_{ij}(s)$ of the matrix $\mathbb{B}(s)$ and then compute all the Laplace transforms $P^*_{ij}(s) = B_{ji}(s)/\Delta(s)$.

Step 3. Expand $P^*_{ij}(s)$ in partial fractions

$$P^*_{ij}(s) = \sum_{k=0}^{n-1} \alpha_{ij}(k)/(s - s_k),$$

namely, compute the $\alpha_{ij}(k)$ via

$$\alpha_{ij}(k) = (s - s_k)P_{ij}^*(s)|_{s=s_k}.$$

Step 4 (the final step). Invert $P_{ij}^*(s)$ to

$$P_{ij}(t) = \alpha_{ij}(0) + \sum_{k=1}^{n-1} \alpha_{ij}(k)e^{s_k t}. \qquad (6.11)$$

Since all $s_k < 0$ $(k \geq 1)$, the $\alpha_{ij}(0) := p_j$ are the steady-state probabilities (to be discussed in the next section).

Eigen vectors method. This method is based on two facts. One is that if V and D are both $n \times n$ matrices, then

$$Q = V^{-1}DV \quad \Leftrightarrow \quad e^Q = V^{-1}e^D V.$$

The second fact is that if D is a diagonal matrix with diagonal elements d_k, then e^D is also diagonal,

$$e^D = \begin{pmatrix} e^{d_0} & 0 & \cdots & 0 \\ 0 & e^{d_1} & & \vdots \\ \vdots & & \ddots & 0 \\ 0 & \cdots & 0 & e^{d_{n-1}} \end{pmatrix}.$$

To evaluate $\mathbb{P}(t)$, proceed as follows.

Step 1. Compute the characteristic polynomial $\Delta(s)$ of Q and its roots $s_0 = 0, s_1, \ldots, s_{n-1}$, known as the *eigen values*.

Step 2. For each eigen value s_k, compute its *left eigen vector* $\mathbf{v}_k \in \mathbb{R}^n$, namely a row-vector which solves

$$\mathbf{v}_k Q = s_k \mathbf{v}_k \quad \Leftrightarrow \quad \mathbf{v}_k \mathbb{B}(s_k) = \mathbf{0}.$$

Step 3. Define the matrix V whose kth row is $\mathbf{v}_k$ and compute its inverse V^{-1}. (Verify that $Q = V^{-1}DV$, where $D = \text{diag}(s_0, s_1, \ldots, s_{n-1})$ and that the columns of V^{-1} are right eigen vectors of Q.)

Step 4 (the final step). Compute the product

$$V^{-1} \begin{pmatrix} e^{s_0 t} & 0 & \cdots & 0 \\ 0 & e^{s_1 t} & & \vdots \\ \vdots & & \ddots & 0 \\ 0 & \cdots & 0 & e^{s_{n-1} t} \end{pmatrix} V = e^{Qt} = \mathbb{P}(t).$$

Taylor expansion method. A method which avoids the computation of eigen vectors and the inverse of an $n \times n$ matrix V is the Taylor expansion. This method is based on Equation (6.5), namely

$$\mathbb{P}(t) = I + tQ + t^2 Q^2 / 2! + \cdots, \qquad (6.12)$$

which is convenient if all the powers Q^m of Q are easily derived in closed form (as in the numerical example). However, computing only the first few terms of the expansion for a particular t might be a good approximation. In general, computing the first $n - 1$ powers of the matrix Q is sufficient for the full evaluation of the transition probabilities matrix $\mathbb{P}(t)$. From Equation 6.11 we know that

$$P_{ij}(t) = \sum_{k=0}^{n-1} \alpha_{ij}(k) e^{s_k t} = \sum_{m=0}^{\infty} \left(\sum_{k=0}^{n-1} \alpha_{ij}(k) s_k^m \right) \frac{t^m}{m!}.$$

Hence,

$$\sum_{k=0}^{n-1} \alpha_{ij}(k) s_k^m = (Q^m)_{ij},$$

where $(Q^m)_{ij}$ is the i, j element of the matrix Q^m. To evaluate $\mathbb{P}(t)$, proceed as follows.

Step 1. Compute the characteristic polynomial $\Delta(s)$ of Q and its roots $s_0 = 0, s_1, \ldots, s_{n-1}$.

Step 2. Compute $Q^2, Q^3, \ldots, Q^{n-1}$.

Step 3. For every pair i, j solve the n linear equations with the unknowns $\alpha_{ij}(0), \alpha_{ij}(1), \ldots, \alpha_{ij}(n-1)$:

$$\sum_{k=0}^{n-1} \alpha_{ij}(k) s_k^m = (Q^m)_{ij}, \quad m = 0, 1, \ldots, n-1.$$

Step 4 (the final step). Use the solution for the α_{ij} to compute

$$P_{ij}(t) = \sum_{k=0}^{n-1} \alpha_{ij}(k) e^{s_k t}.$$

6.1.4 A numerical example (continued)

Looking back at Model 2(a) (two identical active units, one repairman) with failure rate $\lambda = 1$ and repair rate $\mu = 6$, we first draw the transitions and rates diagram as appears in Figure 6.2.

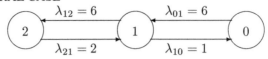

Figure 6.2 *Transitions and rates diagram of Model 2(a), two identical active units and one re-pairman.*

The transition rates matrix Q is given by

$$Q = \begin{pmatrix} -6 & 6 & 0 \\ 1 & -7 & 6 \\ 0 & 2 & -2 \end{pmatrix}.$$

Since we have already applied the Laplace method to this example, we will only apply the other two.

Eigen vectors method. We have already found the eigen values of Q to be $s_0 = 0$, $s_1 = -5$, $s_2 = -10$. The corresponding left eigen vectors (found via $\mathbf{v}_k Q = s_k \mathbf{v}_k$) are given as the rows of the matrix V,

$$V = \begin{pmatrix} 1 & 6 & 18 \\ 1 & 1 & -2 \\ 1 & -4 & 3 \end{pmatrix}.$$

The inverse is found to be

$$V^{-1} = \frac{1}{25} \begin{pmatrix} 1 & 18 & 6 \\ 1 & 3 & -4 \\ 1 & -2 & 1 \end{pmatrix}.$$

The final result is obtained from the product of three matrices

$$\mathbb{P}(t) = e^{Qt} = V^{-1} \begin{pmatrix} 1 & 0 & 0 \\ 0 & e^{-5t} & 0 \\ 0 & 0 & e^{-10t} \end{pmatrix} V. \tag{6.13}$$

We leave it to the reader to carry out the multiplication in Equation (6.13) and see that all nine entries agree with the ones obtained in Subsection 5.2.2.

We remind the reader that an eigen vector is unique up to a scale factor, and we can choose a convenient scale (e.g., such that the coordinates of the eigen vectors are integers). The final result is scale-free, since V and V^{-1} both appear in the product (6.13).

Taylor expansion. One can show by induction (or otherwise) that for all integer

$m \geq 1$,

$$Q^m = (-1)^m 5^{m-2} \begin{pmatrix} 6(2^m+3) & 6(3-2^{m+2}) & 18(2^m-2) \\ 3-2^{m+2} & 2^{m+4}+3 & -6(2^{m+1}+1) \\ 2^m-2 & -2^{m+2}-2 & 3(2^m)+4 \end{pmatrix}.$$

To compute $\mathbb{P}(t) = e^{Qt} = I + \Sigma_1^\infty Q^m t^m/m!$, we carry out the summation entry by entry. For instance,

$$\begin{aligned} P_{00}(t) &= 1 + \sum_{m=1}^\infty (-1)^m 5^{m-2} 6(2^m+3) t^m/m! \\ &= 1 + \tfrac{6}{25}\sum_{m=1}^\infty (-10t)^m/m! + \tfrac{18}{25}\sum_{m=1}^\infty (-5t)^m/m! \\ &= 1 + \tfrac{6}{25}(e^{-10t}-1) + \tfrac{18}{25}(e^{-5t}-1) \\ &= \tfrac{1}{25} + \tfrac{18}{25}e^{-5t} + \tfrac{6}{25}e^{-10t}. \end{aligned}$$

Similarly, for an entry off the main diagonal,

$$\begin{aligned} P_{10}(t) &= \sum_{m=1}^\infty (-1)^m 5^{m-2}(3-2^{m+2}) t^m/m! \\ &= \tfrac{3}{25}\sum_{m=1}^\infty (-5t)^m/m! - \tfrac{4}{25}\sum_{m=1}^\infty (-10t)^m/m! \\ &= \tfrac{3}{25}(e^{-5t}-1) - \tfrac{4}{25}(e^{-10t}-1) \\ &= \tfrac{1}{25} + \tfrac{3}{25}e^{-5t} - \tfrac{4}{25}e^{-10t}. \end{aligned}$$

For the other version of this method, we need $Q^0 = I$, Q and

$$Q^2 = \begin{pmatrix} 42 & -78 & 36 \\ -13 & 67 & -54 \\ 2 & -18 & 16 \end{pmatrix}.$$

Recall, the characteristic roots are $s_0 = 0$, $s_1 = -5$, $s_2 = -10$. Suppose we want to compute $P_{00}(t)$ first. Then we have to solve the three equations:

$$\begin{array}{rcrcrcl} \alpha_{00}(0) &+& \alpha_{00}(1) &+& \alpha_{00}(2) &=& 1 \\ & & -5\alpha_{00}(1) &+& -10\alpha_{00}(2) &=& -6 \\ & & 25\alpha_{00}(1) &+& 100\alpha_{00}(2) &=& 42 \end{array}.$$

The solution is $(1,\ 18,\ 6)/25$, which gives $P_{00}(t)$ as above. For computing $P_{10}(t)$ we have to solve

$$
\begin{aligned}
\alpha_{10}(0) \quad + \quad \alpha_{10}(1) \quad + \quad \alpha_{10}(2) &= 0 \\
-5\alpha_{10}(1) \quad + \quad -10\alpha_{10}(2) &= 1 \\
25\alpha_{10}(1) \quad + \quad 100\alpha_{10}(2) &= -13
\end{aligned}
$$

The solution is $(1, \ 3, \ -4)/25$, which gives $P_{10}(t)$ as above. The other seven entries of $\mathbb{P}(t)$ are solved similarly.

Remark 1 Nowadays, there exist a number of easily accessible mathematical softwares which can carry out these computations very conveniently. Here are two examples:

1. *Mathematica*

```
In[1]:= Q={{-6, 6, 0},{1, -7, 6},{0, 2, -2}};

In[2]:= P[t_]=MatrixExp[Q*t]

            5 t      10 t              5 t      10 t                    5 t 2
   6 + 18 E     + E        6 (-4 + 3 E     + E     )    18 (-1 + E      )
={{-------------------,   ------------------------,   ----------------},
            10 t                      10 t                      10 t
       25 E                       25 E                      25 E


           5 t     10 t           5 t      10 t            5 t       10 t
   -4 + 3 E    + E        16 + 3 E     + 6 E        6 (-2 - E     + 3 E)
   {-------------------,   ----------------------,   -----------------------},
            10 t                      10 t                      10 t
       25 E                       25 E                      25 E


        5 t 2           5 t      10 t           5 t      10 t
   (-1 + E   )    2 (-2 - E     + 3 E     )   3 + 4 E     + 18 E
   {-------------,   ----------------------,   ---------------------}
        10 t                      10 t                      10 t
   25 E                       25 E                      25 E
```

2. MATLAB

```
>> Q=[-6 6 0; 1 -7 6; 0 2 -2];
>> syms t
>> P=expm(Q*t)

P =

[1/25+18/25*exp(-5*t)+6/25*exp(-10*t),
    -24/25*exp(-10*t)+18/25*exp(-5*t)+6/25,
        18/25*exp(-10*t)-36/25*exp(-5*t)+18/25]

[3/25*exp(-5*t)+1/25-4/25*exp(-10*t),
    6/25+16/25*exp(-10*t)+3/25*exp(-5*t),
        -12/25*exp(-10*t)-6/25*exp(-5*t)+18/25]
```

```
[1/25*exp(-10*t)-2/25*exp(-5*t)+1/25,
     -2/25*exp(-5*t)+6/25-4/25*exp(-10*t),
          18/25+3/25*exp(-10*t)+4/25*exp(-5*t)]
```

6.1.5 Multiplicity of roots

Until now, it was implicitly assumed that all the eigen values are distinct. If there exist eigen values with multiplicity greater than 1, then one has to compute the Jordan canonical form of Q, J, say, $Q = T^{-1}JT$ with the proper matrix T, and then $e^{Qt} = T^{-1}e^{Jt}T$. This implies that some of the terms e^{s_it} are multiplied by a power of t. More specifically, if, say, $s_1 = s_2 = \cdots = s_m$, then $e^{s_2t}, \ldots, e^{s_mt}$ are replaced by $te^{s_1t}, \ldots, t^{m-1}e^{s_1t}$, respectively. Here is an example. Suppose we have a one-unit system with failure rate $\lambda_1 = 1$. When the unit fails, a call for service is sent immediately, but it takes an exponential amount of time with rate $\lambda_2 = 4$ until the repair starts. The repair rate is $\mu = 9$. Let E_1 be the operational state, E_w the waiting for service state and E_0 the repair state. Then the transition rates matrix Q is given by

$$Q = \begin{pmatrix} -1 & 1 & 0 \\ 0 & -4 & 4 \\ 9 & 0 & -9 \end{pmatrix}.$$

The characteristic polynomial $\Delta(s) = |I - Q| = s(s+7)^2$, hence $s_0 = 0$, $s_1 = s_2 = -7$. Using the Laplace transform method,

$$P_{11}^*(s) = \frac{1}{\Delta(s)} \begin{vmatrix} 1 & 0 & 0 \\ 0 & s+4 & -4 \\ -9 & 0 & s+9 \end{vmatrix}$$

$$= \frac{(s+4)(s+9)}{\Delta(s)} = \frac{36/49}{s} + \frac{13/49}{s+7} + \frac{42/49}{(s+7)^2}$$

implying

$$P_{11}(t) = \frac{36}{49} + \frac{13}{49}e^{-7t} + \frac{42}{49}te^{-7t}.$$

Using the notation of Subsection 5.2.2, let $\mathbf{y}(t) = (1, e^{-7t}, te^{-7t})$. Then $\mathbf{P}_i(t) = \mathbf{y}(t)\mathbb{C}_i$ for $i = 1, w, 0$, where

$$\mathbb{C}_1 = \frac{1}{49}\begin{pmatrix} 36 & 9 & 4 \\ 13 & -9 & -4 \\ 42 & -14 & -28 \end{pmatrix}$$

$$\mathbb{C}_w = \frac{1}{49}\begin{pmatrix} 36 & 9 & 4 \\ -36 & 40 & -4 \\ -252 & 84 & 168 \end{pmatrix}$$

$$\mathbb{C}_0 = \frac{1}{49}\begin{pmatrix} 36 & 9 & 4 \\ -36 & -9 & 45 \\ 189 & -63 & -126 \end{pmatrix}.$$

6.1.6 Steady-state analysis

We are dealing here with irreducible systems (i.e., $q_i > 0$ for all states i) and a finite state space. For these systems, the limit of $P_{ij}(t)$, as $t \to \infty$, exists for all i, j and is independent of the initial state i, $\lim_{t\to\infty} P_{ij}(t) = p_j > 0$. Let $\boldsymbol{\pi} = (p_0, p_1, \ldots)$ be the row-vector of the limiting distribution, then it follows that the matrix $\mathbb{P}(t)$ converges to $\boldsymbol{\Pi}$, a matrix whose rows are all equal to $\boldsymbol{\pi}$. This in turn implies that

$$\lim_{t\to\infty} \boldsymbol{\pi}_0 \mathbb{P}(t) = \boldsymbol{\pi} \quad \text{for all initial distributions } \boldsymbol{\pi}_0, \tag{6.14}$$

$$\boldsymbol{\pi}\mathbb{P}(t) = \boldsymbol{\pi} \quad \text{for all } t \geq 0 \tag{6.15}$$

and

$$\boldsymbol{\pi} Q = \mathbf{0}. \tag{6.16}$$

The identity (6.15) follows from Equation (6.14), applied to

$$\lim_{s\to\infty} \boldsymbol{\pi}\mathbb{P}(s + t) = \boldsymbol{\pi} = (\lim_{s\to\infty} \boldsymbol{\pi}\mathbb{P}(s))\mathbb{P}(t) = \boldsymbol{\pi}\mathbb{P}(t).$$

Equation (6.16) is just the derivative of Equation (6.15), evaluated at $t = 0$.

While Equation (6.14) means that $\boldsymbol{\pi}$ is the limiting distribution of the system over the possible states, Equation (6.15) means that $\boldsymbol{\pi}$ is the *stationary* or *steady-state* distribution. When such a distribution exists the system is said to be *ergodic* and

$$\lim_{T\to\infty} \frac{1}{T} \int_0^T \mathbf{1}\{\xi(t) = j\}dt = p_j. \tag{6.17}$$

This means that the fraction of time the system spends in state j tends to p_j. Similarly, if $N_{ij} = N_{ij}(T)$ is the number of (direct) transitions from state i to state j during $[0, T]$, $N_i = \sum_{j \neq i} N_{ij}$ is the total number of visits to state i and $N = \sum_i N_i$ is the total number of transitions, then the following limits exist (or, laws of large numbers hold):

$$\lim_{T \to \infty} \frac{N_{ij}}{T} = p_i \lambda_{ij},$$

$$\lim_{T \to \infty} \frac{N_i}{T} = p_i q_i,$$

$$\lim_{T \to \infty} \frac{N}{T} = \sum_i p_i q_i,$$

$$\lim_{N \to \infty} \frac{N_i}{N} = \frac{p_i q_i}{\sum_j p_j q_j}, \qquad (6.18)$$

$$\lim_{N \to \infty} \frac{N_{ij}}{N_i} = p_{ij},$$

$$\lim_{N \to \infty} \frac{N_{ij}}{N} = \frac{p_i q_i p_{ij}}{\sum_j p_j q_j}.$$

Given a continuous-time Markov chain with all its transitions rates λ_{ij}, it is useful to draw the transitions and rates diagram and compute the infinitesimal generator matrix Q. If we are only interested in the stationary distribution π (and not $\mathbb{P}(t)$), then we just have to solve Equation (6.16), subject to the fact that π is a distribution over the state space, that is $\sum_0^{n-1} p_j = 1$. Recalling that $q_{ij} = \lambda_{ij}$ for $i \neq j$ and $q_{ii} = -q_i = -\sum_{j \neq i} q_{ij}$, we have to solve

$$\sum_{j \neq i} p_j \lambda_{ji} - p_i q_i = 0, \quad i = 0, 1, \ldots, n-1. \qquad (6.19)$$

These equations express the *steady-state equilibrium* in each state— the input is equal to the output.

In the numerical example of Subsection 5.2.2, one can check that each row of $\mathbb{P}(t)$ converges to $\pi = \frac{1}{25}(1, 6, 18)$ and that $\pi \mathbb{P}(t) = \pi$ for all $t \geq 0$.

6.2 Reliability of three-unit repairable systems

6.2.1 Steady-state analysis

We consider the following three-unit repairable systems, enumerated in Table 6.1.

Assume that the times to failure of a single unit are exponentially distributed with p.d.f. $f(t) = \lambda e^{-\lambda t}$, $t \geq 0$ and that the time to repair a unit is expo-nentially distributed with p.d.f. $g(t) = \mu e^{-\mu t}$, $t \geq 0$. Let E_j denote the state in which j units are good and $3 - j$ units are failed. Recalling that $P_{ij}(t)$ is the probability of being in state E_j at time t, given that the system is initially

Table 6.1

Case	No. of Active Units	No. of Inac. Units	No. of Repairmen
3(a)	3	0	1
3(b)	2	1	1
3(c)	1	2	1
3(d)	3	0	2
3(e)	2	1	2
3(f)	1	2	2
3(g)	3	0	3
3(h)	2	1	3
3(i)	1	2	3

in state E_i. The steady state probabilities $p_j = \lim_{t \to \infty} P_{ij}(t)$ exist independently of the initial state E_i. Letting $x = \mu/\lambda$, Table 6.2 gives the formulae for p_0, p_1, p_2, and p_3 for each of the cases of Table 6.1.

Table 6.2

Case	p_3	p_2	p_1	p_0
3(a)	$\frac{x^3}{3!}p_0$	$\frac{x^2}{2!}p_0$	xp_0	$\{1 + x + \frac{x^2}{2!} + \frac{x^3}{3!}\}^{-1}$
3(b)	$\frac{x^3}{4}p_0$	$\frac{x^2}{2}p_0$	xp_0	$\{1 + x + \frac{x^2}{2} + \frac{x^3}{4}\}^{-1}$
3(c)	$x^3 p_0$	$x^2 p_0$	xp_0	$\{1 + x + x^2 + x^3\}^{-1}$
3(d)	$\frac{2x^3}{3}p_0$	$2x^2 p_0$	$2xp_0$	$\{1 + 2x + 2x^2 + \frac{2x^3}{3}\}^{-1}$
3(e)	$x^3 p_0$	$2x^2 p_0$	$2xp_0$	$\{1 + 2x + 2x^2 + x^3\}^{-1}$
3(f)	$4x^3 p_0$	$4x^2 p_0$	$2xp_0$	$\{1 + 2x + 4x^2 + 4x^3\}^{-1}$
3(g)	$x^3 p_0$	$3x^2 p_0$	$3xp_0$	$\{1 + x\}^{-3}$
3(h)	$\frac{3x^3}{2}p_0$	$3x^2 p_0$	$3xp_0$	$\{1 + 3x + 3x^2 + \frac{3x^3}{2}\}^{-1}$
3(i)	$6x^3 p_0$	$6x^2 p_0$	$3xp_0$	$\{1 + 3x + 6x^2 + 6x^3\}^{-1}$

If a system is considered to be in a good state when in E_1, E_2, or E_3, and in a bad state when in E_0, then the system unavailability $= p_0$ and the system availability $= 1 - p_0$.

6.3 Steady-state results for the n-unit repairable system

In this section, we generalize the steady-state results of Chapter 5 and Section 6.1. Assume that there are $n+1$ possible states $E_0, E_1, E_2, \ldots, E_n$ where E_j is the state in which j units are in a good state. Let us assume further that the event involving a transition from one state to another state in the short time interval $(t, t + \Delta t]$ is independent of the past and that the (conditional) transition probabilities from one state to another in $(t, t + \Delta t]$ are as given in Table 6.3.

Table 6.3

Transition	Probability of transition in $(t, t + \Delta t]$
$E_n \to E_n$	$1 - \lambda_n \Delta t + o(\Delta t)$
$E_n \to E_{n-1}$	$\lambda_n \Delta t + o(\Delta t)$
$E_j \to E_{j+1}$	$\mu_j \Delta t + o(\Delta t)$
$E_j \to E_j$ if $j = 1, 2, \ldots, n - 1$	$1 - (\lambda_j + \mu_j)\Delta t + o(\Delta t)$
$E_j \to E_{j-1}$	$\lambda_j \Delta t + o(\Delta t)$
$E_0 \to E_1$	$\mu_0 \Delta t + o(\Delta t)$
$E_0 \to E_0$	$1 - \mu_0 \Delta t + o(\Delta t)$
all other transitions	$o(\Delta t)$

Proceeding in the usual way, we can set up the basic difference equations which relate the probability of being in a certain state at time $t + \Delta t$ to the probabilities of being in various states at time t. From these difference equations, we obtain a system of differential equations. Letting $t \to \infty$, we obtain a system of algebraic equations for the steady state probabilities $p_i = \lim_{t \to \infty} P_i(t)$.

These equations are:

$$\lambda_1 p_1 - \mu_0 p_0 = 0$$
$$\lambda_{j+1} p_{j+1} - (\lambda_j + \mu_j)p_j + \mu_{j-1} p_{j-1} = 0, \quad j = 1, 2, \ldots, n - 1$$
$$\lambda_n p_n - \mu_{n-1} p_{n-1} = 0. \tag{6.20}$$

In addition, we have the condition $p_0 + p_1 + \cdots + p_n = 1$. It is easy to verify that the solution to the system of Equations (6.20) is given by

$$p_{j+1} = \frac{\mu_j}{\lambda_{j+1}} p_j = \frac{\mu_0 \mu_1 \cdots \mu_j}{\lambda_1 \lambda_2 \cdots \lambda_{j+1}} p_0, \quad j = 0, 1, 2, \ldots, n - 1, \tag{6.21}$$

where

$$p_0 = \frac{1}{1 + \frac{\mu_0}{\lambda_1} + \frac{\mu_0 \mu_1}{\lambda_1 \lambda_2} + \cdots + \frac{\mu_0 \mu_1 \cdots \mu_{n-1}}{\lambda_1 \lambda_2 \ldots \lambda_n}},$$

since $\sum_{i=0}^{n} p_i = 1$. In terms of the λ_{ij}, here $\lambda_{i,i-1} = \lambda_i$ and $\lambda_{i-1,i} = \mu_{i-1}$ for $i = 1, 2, \ldots, n$. If $|i - j| > 1$ then $\lambda_{ij} = 0$.

If $E_1, E_2, \ldots, E_n$ are considered to be good states for the system and E_0

is a bad state, then p_0 is the (long-run) system unavailability. More generally, if $E_r, E_{r+1}, \ldots, E_n$ are considered to be good states for the system and $E_0, E_1, \ldots, E_{r-1}$ are considered to be bad states, then $U = \sum_{j=0}^{r-1} p_j$ is the (long-run) system unavailability. Here are some examples.

6.3.1 Example 1 — Case 3(e)

Suppose that we have three units—two active, one inactive— and two repair facilities. Assume that the failure rate of an active unit is λ and that the failure rate of an inactive unit is zero until called upon (when it becomes λ). Each repair facility has repair rate μ.

Let E_j, $j = 0, 1, 2, 3$, be the state in which j units are good. For this example $\mu_0 = 2\mu$, $\mu_1 = 2\mu$, $\mu_2 = \mu$; $\lambda_1 = \lambda$, $\lambda_2 = 2\lambda$, $\lambda_3 = 2\lambda$. Hence, substituting in Equation (6.21) and letting $x = \mu/\lambda$, it follows that $p_1 = 2xp_0$, $p_2 = 2x^2p_0$, $p_3 = x^3p_0$, where $p_0 = 1/(1 + 2x + 2x^2 + x^3)$.

6.3.2 Example 2

Suppose that we have n units, all active, each with failure rate λ and one repair facility with repair rate μ. For this case, $\mu_0 = \mu_1 = \mu_2 = \cdots = \mu_{n-1} = \mu$ and $\lambda_j = j\lambda$, for $j = 1, 2, \ldots, n$. If we let $x = \mu/\lambda$ and substitute in Equation (6.21), we obtain

$$p_j = \frac{x^j}{j!} p_0, \qquad j = 1, 2, \ldots, n, \qquad \text{where}$$

$$p_0 = \frac{1}{\sum_{j=0}^{n} x^j/j!}.$$

Hence $p_j = p(j; x)/\pi(n; x)$, where

$$p(j; x) = e^{-x}x^j/j! \quad \text{and} \quad \pi(n; x) = \sum_{j=0}^{n} e^{-x}x^j/j!.$$

Comment: If the time to unit failure is distributed with c.d.f. $F(t)$ possessing finite mean $\alpha = \int_0^\infty t\,dF(t)$, then these formulae hold with $x = \alpha\mu$.

6.3.3 Example 3

Suppose that we have n units all active, each with failure rate λ, and n independent repair facilities, each with repair rate μ. For this case $\mu_j = (n-j)\mu$, for $j = 0, 1, 2, \ldots, n - 1$ and $\lambda_j = j\lambda$, for $j = 1, 2, \ldots, n$. If we let $x = \mu/\lambda$ and substitute in Equation (6.21), we obtain

$$p_j = \frac{n(n-1)\ldots(n-j+1)x^j}{1,2,3,\ldots,j}\, p_0 = \binom{n}{j} x^j p_0, \quad j = 1, 2, \ldots, n$$

where

$$p_0 = \frac{1}{\sum_{j=0}^{n} \binom{n}{j} x^j} = \frac{1}{(1+x)^n}.$$

Hence

$$p_j = \binom{n}{j} \left(\frac{x}{1+x}\right)^j \left(\frac{1}{1+x}\right)^{n-j}, \quad j = 0, 1, 2, \ldots, n.$$

This is the binomial distribution and is written as $p_j = b(j; n, x/(1+x))$, the probability of exactly j successes in n Bernoulli trials and the probability of success on each trial $= x/(1+x)$ (and the probability of failure on each trial $= 1/(1+x)$). If the system is considered to be in a good state when at least k out of the n units are in a good state, then the system availability is given by

$$\sum_{j=k}^{n} p_j = \sum_{j=k}^{n} b\left(j; n, \frac{x}{1+x}\right) = 1 - B\left(k-1; n, \frac{x}{1+x}\right),$$

where

$$B\left(k-1; n, \frac{x}{1+x}\right) = \sum_{j=0}^{k-1} b\left(j; n, \frac{x}{1+x}\right)$$

is the probability of observing $(k-1)$ or fewer successes in n Bernoulli trials, and the probability of success on each trial is $x/(1+x)$.

6.3.4 Example 4

Suppose that we have n units— one active and $(n-1)$ inactive and one repairman. An active unit has failure rate λ; an inactive unit has failure rate 0 until it is placed in service. The repair rate of a failed unit is μ. For this case $\mu_j = \mu$, $j = 0, 1, 2, \ldots, n-1$ and $\lambda_j = \lambda$, $j = 1, 2, \ldots, n$. If we let $x = \mu/\lambda$ and substitute in Equation (6.21), we get $p_j = x^j p_0$, $j = 1, 2, \ldots, n$, where

$$p_0 = \frac{1}{1 + x + x^2 + \cdots + x^n} = \frac{x-1}{x^{n+1}-1}.$$

Hence

$$p_j = \frac{x^{j+1} - x^j}{x^{n+1} - 1}, \quad j = 0, 1, 2, \ldots, n.$$

6.4 Pure birth and death processes

A stochastic process whose transition probabilities satisfy Table 6.3 is called a *birth and death process*. An excellent discussion of this process is given in Chapter 17 of Feller [26]. An important special case of the birth and death process, which results from assigning appropriate values to λ_j and μ_j, is the machine repairman system. For the special case where $\mu_j = 0$ for all j, we have a pure death process. An example of a pure death process is a machine repairman system with no repairmen, i.e., the nonrepairable case. The only transitions possible in a pure death process are $E_n \rightarrow E_{n-1} \rightarrow E_{n-2} \rightarrow \cdots \rightarrow E_0$. Here are some examples.

6.4.1 Example 1

Consider a system consisting of n units all active, where the life time of each unit has the common p.d.f. $f(t) = \lambda e^{-\lambda t}$, $t \geq 0$. In this case, we have a pure death process with $\lambda_j = j\lambda$, for $j = 1, 2, \ldots, n$. Assuming that the system is initially in E_n (all units are good), it can be shown that

$$P_j(t) = \binom{n}{j} \left(1 - e^{-\lambda t}\right)^{n-j} e^{-j\lambda t}, \quad j = 0, 1, 2, \ldots, n. \tag{6.22}$$

It also can be verified that $p_j = \lim_{t\to\infty} P_j(t) = 0$, for $j = 1, 2, \ldots, n$ and $p_0 = \lim_{t\to\infty} P_0(t) = 1$. Another way of saying this is that E_0 (all items have failed) is an absorbing state and all other states $E_1, E_2, \ldots, E_n$ are transient states. Hence $P_0(t)$ is the c.d.f. of the time to reach state E_0, the failed state.

6.4.2 Example 2

Consider a system consisting of one active unit and $(n-1)$ inactive standby units. The active unit has p.d.f. $f(t) = \lambda e^{-\lambda t}$, $t \geq 0$. Standby units are called into service one-by-one until they are all used up. Prior to service, the failure rate of a standby unit equals zero and then becomes λ when called into service. In this case, we have a pure death process with $\lambda_j = \lambda$, for $j = 1, 2, \ldots, n$. Assuming that the system is initially in E_n (all units good), it can be shown that

$$P_j(t) = \frac{(\lambda t)^{n-j}}{(n-j)!} e^{-\lambda t}, \quad j = 1, 2, \ldots, n \tag{6.23}$$

and

$$P_0(t) = 1 - \sum_{k=0}^{n-1} \frac{(\lambda t)^k e^{-\lambda t}}{k!}. \tag{6.24}$$

6.4.3 Example 3

Suppose that we are given k active units and r inactive standby units, which are called into service successively as replacements for units that fail. Each active unit has common life time p.d.f. $f(t) = \lambda e^{-\lambda t}$, $t \geq 0$. Failure is said to occur when the $(r+1)$'st failure occurs. In this case, the initial state is E_{r+k}. The states $E_{r+k}, E_{r+k-1}, \ldots, E_k$ are all transient states and E_{k-1} is an absorbing state. It is readily verified that $\lambda_{r+k} = \lambda_{r+k-1} = \cdots = \lambda_k = k\lambda$.

6.4.4 Example 4

Consider a situation where transitions in a short time interval $(t, t + \Delta t]$ occur with rates as indicated in Figure 6.3.

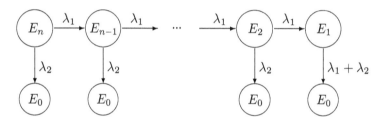

Figure 6.3 *A nonrepairable system with n machines and with a catastrophe.*

Transitions of this type might occur if we have two failure mechanisms operating simultaneously and independently. With one of these failure mechanisms, we reach the absorbing or failed state E_0 in n steps (this could be thought of as wearing out in stages). With the second mechanism, we reach E_0 in a single step (this could be thought of as a catastrophe of some kind). If we let T equal the time to reach E_0, assuming that the system is in E_n at $t = 0$, it is clear that the event $T > t$ occurs if and only if neither failure mechanism causes the system to enter E_0. This means that the event A, "$(n-1)$ or fewer stages in the wear-out process occur during $(0, t]$," and the event B, "there is no catastrophic failure in $(0, t]$," must both occur. Hence

$$P(T > t) = P(A)P(B) = \left[\sum_{j=0}^{n-1} \frac{(\lambda_1 t)^j e^{-\lambda_1 t}}{j!} \right] e^{-\lambda_2 t}$$

$$= e^{-\lambda_2 t} \int_t^\infty \frac{\lambda_1^n u^{n-1} e^{-\lambda_1 u} du}{(n-1)!} . \quad (6.25)$$

It is clear that time-dependent solutions $P_{ij}(t)$, for the n machine-repair system with exponential time to failure and exponential time to repair distributions can be obtained by a direct generalization of the methods given in Chapter 5 for $n = 2$.

The $P_{ij}(t)$ can be obtained by elementary means for the case of n machines and n repairmen. This is done by pairing the men and machines into n independent one-unit repairable systems. Let the time-dependent solutions for the one-unit repairable system be represented as $p_{00}(t)$, $p_{01}(t)$, $p_{10}(t)$, $p_{11}(t)$. It is then easy to verify that $P_{nj}(t)$ is the coefficient of x^j in the binomial expansion $[p_{10}(t) + xp_{11}(t)]^n$, i.e.,

$$P_{nj}(t) = \binom{n}{j} [p_{11}(t)]^j [p_{10}(t)]^{n-j}, \qquad j = 0, 1, 2, \ldots, n. \qquad (6.26)$$

Similarly, $P_{0j}(t)$ is the coefficient of x^j in the binomial expansion of $[p_{00}(t) + xp_{01}(t)]^n$, i.e.,

$$P_{0j}(t) = \binom{n}{j} [p_{01}(t)]^j [p_{00}(t)]^{n-j}, \qquad j = 0, 1, 2, \ldots, n. \qquad (6.27)$$

More generally, $P_{ij}(t)$ is the coefficient of x^j in the expansion of $[p_{10}(t) + xp_{11}(t)]^i [p_{00}(t) + xp_{01}(t)]^{n-i}$, i.e.,

$$P_{ij}(t) = \sum_r \binom{i}{r} \binom{n-i}{j-r} [p_{11}(t)]^r [p_{10}(t)]^{i-r} [p_{01}(t)]^{j-r} [p_{00}(t)]^{n-i-j+r},$$

$$(6.28)$$

where $\max(0, i + j - n) \le r \le \min(i, j)$.

Note that $P_{nj}(t)$ can also be obtained directly as follows: We start with n independent repairable units, each in a good state. At time t, any unit is in a good state with probability $p_{11}(t)$ and in a bad state with probability $p_{10}(t)$. Hence the probability that exactly j units are in a good state at time t is the probability of observing j successes in n Bernoulli trials, with probability of success $p_{11}(t)$ on each trial. This is Equation (6.26). The above formulae are valid for a general (common) unit time to failure and time to repair distribution.

Except for the special case of n machines and n repairmen, there is no analytically convenient time-dependent solution if either the common unit time to failure distribution or common unit time to repair distribution is not exponential. See Chapters 8 and 9 for some steady-state results for the nonexponential machine-repair problem.

6.5 Some statistical considerations

Suppose we observe a continuous time, irreducible, Markov chain during time interval $[0, T]$, where T is large enough to allow all states to be visited several times. In other words, so that the limiting results (6.18) are good approximations. Observing the Markov chain implies that we know its path through the states $E_{j_0} \to E_{j_1} \to E_{j_2} \to \cdots \to E_{j_N}$, where $N = N(T)$ is the number of transitions, and we know the times between transitions $T_0, T_1, \ldots, T_{N-1}$. Note, $T_N = T - \sum_{k=0}^{N-1} T_k$ is the time spent in state E_{j_N} until the observation process is stopped. This random variable is *censored*. Censored observations get special treatment in the statistical literature, but in our case, by our assumptions of large numbers, the censoring effect is negligible. In general, the statistical analysis of the data depends on the initial state and how it is selected. Similarly, the analysis depends on the way the observation process is stopped, i.e., by preassigned T, preassigned N, or any random stopping rule. For instance, if N is preassigned, then every visit at state E_i takes an exponential amount of time with parameter q_i. However, if T is preassigned, then all visits at all states are bounded (by T), thus are not exactly exponentially distributed. Again, if T is large, the assumption that the T_k are independent exponential is a good approximation.

6.5.1 Estimating the rates

Let S_i be the total time spent in state E_i, thus $\sum_i S_i = T$. Let N_i and N_{ij} be as in Subsection 6.1.5. Then, conditioned on the N_i, the MLE and UMVUE of q_i are, respectively,

$$\hat{q}_i = \frac{N_i}{S_i}, \qquad q_i^* = \frac{N_i - 1}{S_i}$$

(see Problem 1.3.6). Out of the N_i visits to state E_i, a proportion $\hat{p}_{ij} = N_{ij}/N_i$ of them terminated with a transition to state E_j. Since $\lambda_{ij} = q_i p_{ij}$, we estimate λ_{ij} by

$$\hat{\lambda}_{ij} = \hat{q}_i \hat{p}_{ij} = \frac{N_i}{S_i} \cdot \frac{N_{ij}}{N_i} = \frac{N_{ij}}{S_i} \quad (i \neq j).$$

It follows that we have estimated all the entries of $Q = (q_{ij})$, the rates matrix of the Markov chain. Namely, if $\hat{Q} = (\hat{q}_{ij})$ is the estimated matrix, then $\hat{q}_{ij} = \hat{\lambda}_{ij}$ for $i \neq j$ and $\hat{q}_{ii} = -\hat{q}_i$. As always, the rates λ_{ij} determine the steady-state probabilities p_i, so the latter can be estimated via the $\hat{\lambda}_{ij}$. However, they can be also estimated directly by

$$\hat{p}_i = \frac{S_i}{T} = \frac{S_i}{\Sigma_k S_k}.$$

It should be noted that all these estimators are consistent (as T or N tend to ∞).

6.5.2 Estimation in a parametric structure

When the rates λ_{ij} are expressible as known functions of a small number of parameters, we can estimate these parameters first and then plug them in those functions to get the rates. This way usually yields more efficient estimates. We will demonstrate the idea through a two-unit repairable system, whose transitions and rates diagram is given in Figure (6.4).

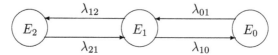

Figure 6.4 *Transitions and rates diagram of a general two-unit repairable system.*

Suppose we have observed the system making $N = 1,000$ transitions and then stopped. The summary of our records is as follows: Total time on test $T = 298$ hours, $S_0 = 12$ hours at state E_0, $S_1 = 72$ hours at state E_1 and $S_2 = 214$ hours at state E_2. The numbers of transitions are $N_{01} = 71$, $N_{10} = 72$, $N_{12} = 428$ and $N_{21} = 429$. Hence, our rates estimates $\hat{\lambda}_{ij} = N_{ij}/S_i$ are

$$\hat{\lambda}_{01} = \frac{71}{12} = 5.91667 \,, \quad \hat{\lambda}_{10} = \frac{72}{72} = 1.0000 \,,$$

$$\hat{\lambda}_{12} = \frac{428}{72} = 5.94444 \,, \quad \hat{\lambda}_{21} = \frac{429}{214} = 2.00467 \,.$$

Now, suppose Figure (6.4) describes Model 2(a). That is, two identical active units, one repairman. Hence we know that $\lambda_{01} = \lambda_{12} = \mu$, $\lambda_{10} = \lambda$ and $\lambda_{21} = 2\lambda$. We have to estimate just μ and λ. An efficient method in this case would be the maximum likelihood. As in Equation (1.30), conditioned on the N_i, the likelihood function is

$$L(\lambda, \mu) = (2\lambda)^{429} e^{-214(2\lambda)} (\lambda + \mu)^{500} e^{-72(\lambda+\mu)} \mu^{71} e^{-12\mu}. \qquad (6.29)$$

We have to find a pair $(\hat{\lambda}, \hat{\mu})$ which maximizes $L(\lambda, \mu)$. For this purpose, we define $\ell = \log L$ and compute

$$\frac{\partial \ell}{\partial \lambda} = \frac{429}{\lambda} + \frac{500}{\lambda + \mu} - 500$$

and

$$\frac{\partial \ell}{\partial \mu} = \frac{71}{\mu} + \frac{500}{\lambda + \mu} - 84 \,.$$

Equating the partial derivatives to 0, we finally have to solve two unknowns from two equations

$$\frac{429}{\lambda} + \frac{500}{\lambda + \mu} = 500$$

and
$$\frac{71}{\mu} + \frac{500}{\lambda + \mu} = 84 \ .$$

After some algebraic manipulations one gets
$$\lambda = \frac{429\mu}{416\mu + 71}$$

and
$$34944\mu^2 - 195536\mu - 71000 = 0 \ .$$

The last two equations yield the MLE of (λ, μ) as $(1.002437, \ 5.937876)$. Finally, for comparison, the MLEs of the λ_{ij} are

$$\hat{\lambda}_{01} = \hat{\lambda}_{12} = \hat{\mu} = 5.93788 \ , \quad \hat{\lambda}_{10} = \hat{\lambda} = 1.00244 \ , \quad \hat{\lambda}_{21} = 2\hat{\lambda} = 2.00488 \ .$$

6.6 Problems and comments

Problems for Section 6.1

1. For models 2(b), 2(c) and 2(d), assume $\lambda = 1$ and $\mu = 6$ and follow the numerical example of Subsection 6.1.4. That is, draw the transitions and rates diagram for each case and determine the infinitesimal generator Q. Then find the matrix of transition probabilities $\mathbb{P}(t)$ (use at least two methods so that you can check the results).

2. A service station consists of two identical computers and two technicians. When both computers are in good condition, most of the work load is on one computer, exposed to a failure rate of $\lambda = 1$, while the other computer's failure rate is $\lambda = .5$. If one of the computers fails, the other one takes the full load, thus exposed to a failure rate of $\lambda = 2$. Among the technicians, one is an expert with repair rate of $\mu = 2$, while the second is a stagiaire, working with repair rate of $\mu = 1$. If both work simultaneously on the same computer, the total repair rate is $\mu = 2.5$. At any given moment, they work so that their total repair rate is maximized.

   (a) Draw the transitions and rates diagram of the system and determine the infinitesimal generator Q.

   (b) Compute the steady-state probabilities and the system availability.

   (c) Compute the transition probabilities matrix $\mathbb{P}(t)$.

3. Consider another service station, similar to the one in Problem 2. Here each computer is exposed to a failure rate of $\lambda = .1$. The expert and the stagiaire work, respectively, with repair rates 2 and 1 as before. However, here only one person at a time can repair a failed computer. If both technicians are free, the expert will fix the next failure, but if the stagiaire has started to repair a failed computer, he will complete his job.

(a) Draw the transitions and rates diagram of the system (note, there are 4 different states) and determine the rates matrix Q.

(b) Compute the steady-state probabilities and the system availability.

(c) Compute the expected number of operational computers at the steady state.

(d) Given that a computer has just failed, what is the (asymptotic) expected time of its repair?

(e) Compute the transition probabilities matrix $\mathbb{P}(t)$.

4. There are two computers in a service station. Computer A fails on the average once per hour and computer B twice per hour. There is one technician on duty who can repair a computer in 3 minutes on average.

(a) Draw the transitions and rates diagram of the system (note, there are 5 different states) and determine the rates matrix Q.

(b) Compute the steady-state probabilities and the system availability. Here, the system is available if at least one computer is good.

(c) Compute the transition probabilities matrix $\mathbb{P}(t)$. This is a greater challenge, to compute a 5×5 matrix. The use of a computing software as Mathematica or MATLAB is recommended.

5. Compute the matrix $\mathbb{P}(t)$ of Subsection 6.1.5.

Problems for Section 6.2

1. Set up tables of transition probabilities from state E_i to state E_j in the short time interval $(t, t + \Delta t]$ for each of the cases 3(a)–3(i).

2. Verify the entries in Table 6.2.

3. Prove in case 3(a) that $p_j = p(j; x)/\pi(3; x)$, $j = 0, 1, 2, 3$ where $p(j; x) = e^{-x}x^j/j!$ and $\pi(3; x) = \sum_{j=0}^{3} p(j; x)$.

4. Compute p_j, $j = 0, 1, 2, 3$ for the case where $x = \mu/\lambda = 5$ for case 3(a).

5. Derive p_0, p_1, p_2 and p_3 in case 3(g) by using the binomial distribution. Hint: For a one-unit repairable system, the probability of being in E_1 is $x/(1 + x)$ and the probability of being in E_0 is $1/(1 + x)$.

6. Suppose that there are three active units and that at least two out of three of these units must be in a good state in order to carry out a successful operation. Graph system availability as a function of $x = \mu/\lambda$ for cases 3(a), (d), (g). For what values of x is the system availability $\geq .99$ in each of the above cases?

Problems for Section 6.3

1. Consider a system consisting of n units—$(n-k)$ active, k inactive, and one repairman, $0 \le k \le n-1$. Assume exponential failure and exponential repair distributions with failure rate λ for active units and failure rate 0 for an inactive unit until it is placed in service. Repair rate is μ. Use Equation (6.21) to find the steady-state probabilities $\{p_i\}$, $i = 0, 1, 2, \ldots, n$.

2. Find the steady-state probabilities $\{p_i\}$, $i = 0, 1, 2, \ldots, n$ for a system composed of n units, all active each with constant failure rate λ, and two repairmen each having constant repair rate μ. It is assumed that only one repairman at a time can work on a failed unit.

3. Find the steady-state probabilities $\{p_i\}$, $i = 0, 1, 2, \ldots, n$ for a system composed of n units, $(n-k)$ active (with constant failure rate λ) and $0 \le k \le n-1$ inactive (with failure rate zero) until called into service, and two repairmen each having constant repair rate μ.

4. Suppose that we have n units and n repairmen. Units fail independently according to a common c.d.f. $F(x)$ possessing finite mean m_x and each repairman has a common time to repair distribution described by the c.d.f. $G(y)$ and possessing finite mean m_y. Let p_j be the long-run probability that j units are in a good state and $(n-j)$ units are in a bad state and under repair. Prove that

$$p_j = \binom{n}{j} \frac{m_x^j m_y^{n-j}}{(m_x + m_y)^n} .$$

Hint: Pair off the n units and n repairmen into n independent machine–repairman combinations, i.e., as n independent one-unit repairable systems. Then use the binomial theorem.

5. Suppose that we have n units—one active and $(n-1)$ inactive— and n repairmen. An active unit has failure rate λ; an inactive unit has failure rate 0 until it is placed in service. Only one repairman can work on a single failed unit at a time. The repair rate on a failed unit is μ. Let p_j be the long-run probability that j units are in a good state (1 active, $(j-1)$ inactive but good) and $(n-j)$ units are in a bad state and under repair. Prove that

$$p_j = \frac{n^{n-j}}{(n-j)!} \bigg/ \sum_{k=0}^{n} \frac{y^{n-k}}{(n-k)!} = \frac{p(n-j;y)}{\pi(n;y)} ,$$

where $y = \lambda/\mu$.

Comment. It is of interest that if the time to repair a unit is distributed with general c.d.f. $G(t)$ with mean $\beta = \int_0^\infty t\, dG(t)$, then the above result holds with $y = \lambda\beta$.

6. Suppose that E_0 (all units are failed) is the "bad" state for the system in examples 2 and 4. Compute the long-run system availability for each of these examples if $x = \mu/\lambda = 3$ and $n = 4$.

Problems for Section 6.4

1. Obtain Equation (6.3) directly and also by setting up a system of differential equations and then solving by using Laplace transforms.

2. Derive Equations (6.23) and (6.24) directly and also by setting up a system of differential equations and solving by Laplace transforms. Verify that E_i, $i = 1, 2, \ldots, n$ are transient states and that E_0 is an absorbing state.

3. Find the $P_j(t)$ for $j = k - 1, k, \ldots, r + k$ in Example 3.

4. Fill in the gaps in the discussion of Example 4 and derive $P_j(t)$, $j = 1, 2, \ldots, n$.

5. Derive Equation (6.28) by a direct probabilistic argument.

Problem for Section 6.5

1. Consider a pure birth and death process with states 0, 1, 2, and 3. The process was observed making $N = 1,000$ transitions and it was then stopped. It took a total time on test $T = 200$ hours. The records show that the numbers of transitions between the states are as follows: $N_{01} = N_{10} = 14$, $N_{12} = 139$, $N_{21} = 140$, $N_{23} = 346$ and $N_{32} = 347$. The total times spent in the states 0, 1, 2, 3, respectively, are 1.4, 13.9, 69.4 and 115.3 hours.

   (a) Estimate the six rates λ_{ij} of the system.

   (b) Assume the system is Model 3(d), that is, three active units, each with failure rate λ and two repairmen, each with repair rate μ. Estimate the parameters by the maximum likelihood method and compare the results to part (a).

CHAPTER 7

First passage time for systems reliability

7.1 Two-unit repairable systems

7.1.1 Case 2(a) of Section 5.1

The material in this section is based partly on Epstein and Hosford [24].

To start our discussion, let us consider anew the particular system (Case 2(a)) treated in Sections 5.1 and 5.2. This system consists of two identical units, each assumed to be active and possessing a common time to failure p.d.f. $f(t) = \lambda \exp\{-\lambda t\}$, $t \geq 0$. It is further assumed that there is a single repairman and that the time to repair a failed unit is exponentially distributed with p.d.f. $g(t) = \mu e^{-\mu t}$, $t \geq 0$. Failure of a unit is assumed to be instantaneously detectable and repair is immediately initiated, if the repairman is free. We let E_j denote the state in which j units are good. In Section 5.2, we found formulae for $P_{ij}(t)$, the probability that the system is in E_j at time t given that it was in E_i at time zero and in Section 5.1 we gave formulae for $p_j = \lim_{t \to \infty} P_{ij}(t)$, the long-run probability of being in state E_j.

In this chapter, we are interested in a different problem. Suppose that the system is initially in E_2 (both units are good). How long will it take for the system to first enter the failed state E_0? This is an example of a *first passage time* problem. Graphically, this means the first time that the random function representing the state of the system as a function of time meets the t-axis (see Figure 7.1).

The method of solution for the first passage time problem for the two-unit system (both active) and one repairman is similar to the methods in Section 5.1. The important technical difference is that E_0 becomes an absorbing state (i.e., transitions from E_0 are not allowed). In terms of the transitions and rates diagram, it means that the arrows exiting from E_0 are erased. In the present case, the new diagram is given in Figure 7.2.

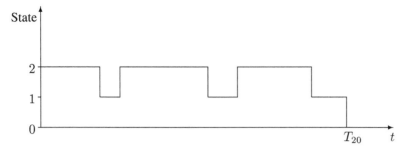

Figure 7.1 *A particular realization of first passage from E_2 into E_0.*

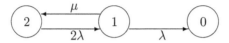

Figure 7.2 *Transitions and rates diagram of Model 2(a), where state E_0 is absorbing.*

The transition rates matrix Q becomes here

$$Q = \begin{pmatrix} 0 & 0 & 0 \\ \lambda & -(\lambda + \mu) & \mu \\ 0 & 2\lambda & -2\lambda \end{pmatrix} \qquad (7.1)$$

(compare to Q in Equation (5.29)).

Since we assume throughout that the initial state is E_2, we write $P_i(t)$ for $P_{2i}(t)$, the probability that the system is in E_i at time t. Then, as before, using the transition rates matrix Q, we have to solve the system of differential equations

$$\begin{aligned} P_2'(t) &= -2\lambda P_2(t) + \mu P_1(t) \\ P_1'(t) &= 2\lambda P_2(t) - (\lambda + \mu)P_1(t) \qquad (7.2) \\ P_0'(t) &= \lambda P_1(t) \,. \end{aligned}$$

The initial conditions are $P_2(0) = 1$, $P_1(0) = 0$, $P_0(0) = 0$.

The steady-state solution of the set of Equations (7.2), which can be obtained normally by setting $P_i'(t) = 0$ and $P_i(t) = p_i$, is of no interest and is trivially given by $p_1 = p_2 = 0$ and $p_0 = 1$, since E_0 is an absorbing state and sooner or later the system reaches E_0. We are particularly interested in $P_0(t)$, the probability of being in E_0 at time t. Since E_0 is an absorbing state, $P_0(t)$ is also the probability of reaching E_0 on or before time t. Thus, if the random variable T is the first passage time into E_0 (the time to first failure of the system starting from E_2), then $P_0(t)$ is the c.d.f. of T, i.e., $P_0(t) = P(T \leq t)$. This c.d.f. is also called *phase-type distribution*. The p.d.f. of time to first failure of the system is given by $P_0'(t)$.

As in Chapter 5, we use the Laplace transform method to obtain $P_0(t)$.

Defining $P_i^*(s)$, the Laplace transforms of $P_i(t)$ as

$$P_i^*(s) = \int_0^\infty e^{-st} P_i(t) dt, \qquad i = 0, 1, 2 \tag{7.3}$$

and applying Laplace transforms to both sides of the set of differential Equations (7.2), we obtain the set of equations

$$
\begin{aligned}
(s + 2\lambda)P_2^*(s) - \mu P_1^*(s) &= 1 \\
2\lambda P_2^*(s) - (\lambda + \mu + s)P_1^*(s) &= 0 \\
sP_0^*(s) - \lambda P_1^*(s) &= 0.
\end{aligned}
\tag{7.4}
$$

Solving for $P_1^*(s)$ from the first two equations of (7.4), we obtain

$$P_1^*(s) = \frac{\begin{vmatrix} s + 2\lambda & 1 \\ 2\lambda & 0 \end{vmatrix}}{\begin{vmatrix} s + 2\lambda & -\mu \\ 2\lambda & -(\lambda + \mu + s) \end{vmatrix}} = \frac{2\lambda}{s^2 + (3\lambda + \mu)s + 2\lambda^2}. \tag{7.5}$$

Solving for the roots s_1 and s_2 of $s^2 + (3\lambda + \mu)s + 2\lambda^2 = 0$, we obtain

$$s_1 = \frac{-(3\lambda + \mu) + \sqrt{(3\lambda + \mu)^2 - 8\lambda^2}}{2} = \frac{-(3\lambda + \mu) + \sqrt{\lambda^2 + 6\lambda\mu + \mu^2}}{2} \tag{7.6}$$

and

$$s_2 = \frac{-(3\lambda + \mu) - \sqrt{\lambda^2 + 6\lambda\mu + \mu^2}}{2}.$$

Hence, expanding in partial fractions,

$$P_1^*(s) = \frac{2\lambda}{(s - s_1)(s - s_2)} = \frac{2\lambda}{s_1 - s_2}\left(\frac{1}{s - s_1} - \frac{1}{s - s_2}\right). \tag{7.7}$$

Using a table of Laplace transforms, it follows that

$$P_1(t) = \frac{2\lambda}{s_1 - s_2}\left(e^{s_1 t} - e^{s_2 t}\right). \tag{7.8}$$

According to the third equation in (7.2), the p.d.f. of T, the time to first failure of the system, is given by $P_0'(t) = \lambda P_1(t)$. Using Equation (7.8), $P_0'(t)$ becomes

$$P_0'(t) = \frac{2\lambda^2}{s_1 - s_2}\left(e^{s_1 t} - e^{s_2 t}\right). \tag{7.9}$$

Integrating both sides of Equation (7.9) and noting that $2\lambda^2 = s_1 s_2$, it is readily verified that the c.d.f. of T is given by

$$P_0(t) = 1 - \frac{s_1 e^{s_2 t} - s_2 e^{s_1 t}}{s_1 - s_2}. \tag{7.10}$$

We define $R(t)$, the reliability of the system at time t, as the probability that the system is not in E_0 at time t (i.e., both units are not down simultaneously anywhere during the time interval $(0, t]$). Therefore,

$$R(t) = 1 - P_0(t) = \frac{s_1 e^{s_2 t} - s_2 e^{s_1 t}}{s_1 - s_2}. \tag{7.11}$$

Note that $R(0) = 1$, which is as it should be since the system is initially in E_2 (both units good). The roots s_1 and s_2 are both negative, and hence $\lim_{t \to \infty} R(t) = 0$. This is also as it should be since E_0 (both units down) is an absorbing state.

Mean time to first system failure It is of particular interest to find the mean time to first system failure (in short MTFSF) (i.e., the mean first passage time from $E_2 \to E_0$). It is easy to verify (see Problem 1 for this section) that the MTFSF is given by

$$\text{MTFSF} = \int_0^\infty R(t)dt = \int_0^\infty \frac{s_1 e^{s_2 t} - s_2 e^{s_1 t}}{s_1 - s_2} dt \tag{7.12}$$

$$= -\left(\frac{1}{s_1} + \frac{1}{s_2} \right) = -\frac{s_1 + s_2}{s_1 s_2} = \frac{3\lambda + \mu}{2\lambda^2}.$$

The last equality is a consequence of the fact that s_1 and s_2 are the roots of the quadratic equation $s^2 + (3\lambda + \mu)s + 2\lambda^2 = 0$. Therefore, the sum of the roots $s_1 + s_2 = -(3\lambda + \mu)$ and the product of the two roots $s_1 s_2 = 2\lambda^2$.

The MTFSF can also be found in the following way. The Laplace transform of $P_0'(t)$, the p.d.f. of the first time to system failure, is given by

$$L\{P_0'\}(s) = \int_0^\infty e^{-st} P_0'(t)dt = \lambda \int_0^\infty e^{-st} P_1(t)dt = \lambda P_1^*(s), \tag{7.13}$$

from Equations (7.2) and (7.3).

Hence, differentiating with respect to s, we have

$$\lambda \frac{d}{ds} P_1^*(s) = -\int_0^\infty t e^{-st} P_0'(t)dt. \tag{7.14}$$

Letting $s = 0$, it follows that

$$\text{MTFSF} = \int_0^\infty t P_0'(t)dt = -\left[\lambda \frac{dP_1^*(s)}{ds} \right]_{s=0}. \tag{7.15}$$

Differentiating Equation (7.5) we have

$$\frac{dP_1^*(s)}{ds} = -2\lambda[2s + (3\lambda + \mu)]/[s^2 + (3\lambda + \mu)s + 2\lambda^2]^2. \qquad (7.16)$$

Hence

$$\frac{dP_1^*(s)}{ds}\bigg|_{s=0} = -\frac{3\lambda + \mu}{2\lambda^3}. \qquad (7.17)$$

Substituting in Equation (7.15) we have

$$\text{MTFSF} = \frac{3\lambda + \mu}{2\lambda^2} \qquad (7.18)$$

which coincides with Equation (7.12).

We shall find it convenient to denote MTFSF as ℓ_{20}, the mean first passage time from E_2 to E_0. More generally, ℓ_{ij} denotes the mean first passage time from state E_i to state E_j. With this notation, we can compute ℓ_{ij} by solving a system of linear equations. Looking back at the transitions and rates diagram in Figure 7.2, we obtain

$$\ell_{10} = \frac{1}{\lambda + \mu} + \frac{\mu}{\lambda + \mu}\ell_{20}$$

$$\ell_{20} = \frac{1}{2\lambda} + \ell_{10}. \qquad (7.19)$$

The first equation follows from the fact that if the system is in state E_1, then the total exit rate is $\lambda+\mu$, implying it will stay there (on the average) $1/(\lambda+\mu)$ amount of time. Then, with probability $\mu/(\lambda + \mu)$ the system will enter state E_2, thus it will need (on the average) ℓ_{20} amount of time to reach state E_0, and with probability $\lambda/(\lambda + \mu)$ it will enter the absorbing state E_0. The argument for the second equation is similar. The solution of the two linear equations is given by

$$\ell_{10} = \frac{2\lambda + \mu}{2\lambda^2}, \qquad \ell_{20} = \frac{3\lambda + \mu}{2\lambda^2},$$

which is in agreement with Equation (7.18).

A comparison with a nonrepairable system. If $\mu = 0$ in the problem treated in this section, we have a nonrepairable system consisting of two active units. It is readily verified that $s_1 = -\lambda$ and $s_2 = -2\lambda$ and substituting in Equation (7.11),

$$R(t) = 1 - P_0(t) = \frac{-\lambda e^{-2\lambda t} + 2\lambda e^{-\lambda t}}{-\lambda + 2\lambda} = 2e^{-\lambda t} - e^{-2\lambda t}$$

$$= 1 - (1 - e^{-\lambda t})^2.$$

This was obtained directly in Section 3.1.

If $\mu = 0$, the mean time to first system failure $= 3/2\lambda$. The mean time to first system failure, when failed units are repaired, is $\ell_{20} = (3\lambda + \mu)/(2\lambda^2)$. The ratio

$$IMP_{2(a)} = \frac{3\lambda + \mu}{2\lambda^2} \bigg/ \frac{3}{2\lambda} = 1 + \frac{\mu}{3\lambda} = 1 + \frac{x}{3},$$

is a measure of the improvement due to repair.

A numerical example (continued). Going back to the numerical example of Subsection 5.2.2, assume the failure rate is $\lambda = 1$ per hour and the repair rate is $\mu = 6$ per hour. Then the matrix Q in this case is given by

$$Q = \begin{pmatrix} 0 & 0 & 0 \\ 1 & -7 & 6 \\ 0 & 2 & -2 \end{pmatrix}.$$

The characteristic polynomial is $\Delta(s) = |sI - Q| = s(s^2 + 9s + 2)$ whose roots are $s_0 = 0$, $s_1 = -.228$ and $s_2 = -8.772$. Substituting in Equation (7.10), the c.d.f. of first time to system failure, given that the system started with two good units, is given by

$$P_0(t) = 1 - R(t) = 1 - 1.02669e^{-.228t} + .02669e^{-8.772t}.$$

Integrating $R(t)$ gives

$$MTFSF = \ell_{20} = \frac{1.02669}{.228} - \frac{.02669}{8.772} = 4.500,$$

which is the same as $(3\lambda + \mu)/(2\lambda^2)$. Here, if $\mu = 0$, then $\ell_{20} = 1.5$ and if $\mu = 6$ then $\ell_{20} = 4.5$. Namely, the improvement factor due to repair is 3.

7.1.2 Case 2(b) of Section 5.1

In our discussion up to this point, we have assumed that both units in the two-unit system are active. We now consider the case where one of the units is an inactive standby (this was called Case 2(b) in Sections 5.1 and 5.2) so that the system being in state E_2 means that both units are operable, but that only one is actually operating, while the other one is standing by and is assumed to be incapable of failure in standby status. When the active unit fails, the standby unit goes into operation and repairs are immediately initiated on the failed unit. If the failed unit is repaired before the currently active unit fails, then it goes into standby status until called into service, etc. It is assumed that the times to failure of an active unit are exponentially distributed with p.d.f. $f(t) = \lambda e^{-\lambda t}$, $t \geq 0$ and that the time to repair is exponentially distributed with p.d.f. $g(t) = \mu e^{-\mu t}$, $t \geq 0$. The transitions and rates diagram is described in Figure 7.3.

The transition rates matrix Q becomes here

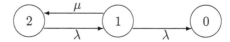

Figure 7.3 *Transitions and rates diagram of Model 2(b), where state E_0 is absorbing.*

$$Q = \begin{pmatrix} 0 & 0 & 0 \\ \lambda & -(\lambda + \mu) & \mu \\ 0 & \lambda & -\lambda \end{pmatrix}. \tag{7.20}$$

The system of differential equations, analogous to Equation (7.2) for Case 2(b) is

$$\begin{aligned} P_2'(t) &= -\lambda P_2(t) + \mu P_1(t) \\ P_1'(t) &= \lambda P_2(t) - (\lambda + \mu) P_1(t) \\ P_0'(t) &= \lambda P_1(t). \end{aligned} \tag{7.21}$$

If the system is assumed to be in E_2 at $t = 0$, it can be verified that

$$P_1^*(s) = \frac{\lambda}{s^2 + (2\lambda + \mu)s + \lambda^2} \tag{7.22}$$

and that

$$R(t) = 1 - P_0(t) = \frac{s_1 e^{s_2 t} - s_2 e^{s_1 t}}{s_1 - s_2}, \tag{7.23}$$

where s_1 and s_2 are the roots of the equation

$$s^2 + (2\lambda + \mu)s + \lambda^2 = 0$$

i.e.,

$$s_1 = \frac{-(2\lambda + \mu) + \sqrt{4\lambda\mu + \mu^2}}{2} \tag{7.24}$$

and

$$s_2 = \frac{-(2\lambda + \mu) - \sqrt{4\lambda\mu + \mu^2}}{2}.$$

The MTFSF ℓ_{20} can be found by solving

$$\ell_{10} = \frac{1}{\lambda + \mu} + \frac{\mu}{\lambda + \mu}\ell_{20} \tag{7.25}$$

$$\ell_{20} = \frac{1}{\lambda} + \ell_{10},$$

the solution of which is

$$\ell_{10} = \frac{\lambda + \mu}{\lambda^2}, \qquad \ell_{20} = \frac{2\lambda + \mu}{\lambda^2}.$$

Alternatively,

$$\ell_{20} = -\frac{d}{ds}\left[\lambda P_1^*(s)\right]_{s=0} = \frac{2\lambda + \mu}{\lambda^2}. \qquad (7.26)$$

A comparison with a nonrepairable system. If $\mu = 0$, which means that we have a nonrepairable system consisting of one active unit and one inactive unit, then Equation (7.22) becomes

$$P_1^*(s) = \lambda/(\lambda + s)^2. \qquad (7.27)$$

Using a table of Laplace transforms, it is readily verified that $P_1^*(s)$ is the Laplace transform of $\lambda t e^{-\lambda t}$. Hence,

$$P_1(t) = \lambda t e^{-\lambda t}. \qquad (7.28)$$

Hence, the p.d.f. of the time to first system failure is given by

$$P_0'(t) = \lambda^2 t e^{-\lambda t}, \quad t \geq 0, \qquad (7.29)$$

which is a gamma density function. Integrating $P_0'(t)$, we obtain the c.d.f. of the time to system failure

$$P_0(t) = 1 - e^{-\lambda t} - \lambda t e^{-\lambda t}. \qquad (7.30)$$

System reliability, $R(t)$, is given by

$$R(t) = 1 - P_0(t) = e^{-\lambda t} + \lambda t e^{-\lambda t}. \qquad (7.31)$$

This was obtained directly in Section 3.4 (see Formula (3.47) with $n = 2$).

If $\mu = 0$, then the mean time to system failure $= 2/\lambda$. The mean time to first system failure when failed units are repaired is given by $\ell_{20} = (2\lambda + \mu)/\lambda^2$. The improvement ratio due to repair here is

$$IMP_{2(b)} = \frac{2\lambda + \mu}{\lambda^2} \Big/ \frac{2}{\lambda} = 1 + \frac{x}{2},$$

which is greater than $IMP_{2(a)} = 1 + x/3$. Furthermore, the MTFSF here is greater than the MTFSF of the 2(a) model, where both units work in parallel. This does not mean we should always prefer model 2(b), since there are other aspects to be considered, e.g., reliability, production rate, etc.

7.2 Repairable systems with three (or more) units

7.2.1 Three units

Consider a three-unit repairable system, where E_i represents the state in which there are i good units. State E_0 is the absorbing state. We will use the notation and assumptions of the pure birth and death processes. Namely, λ_i is the transition rate from state E_i to state E_{i-1} and μ_i is the transition rate from state E_i to state E_{i+1}. All times to birth and to death are exponential with the proper rates.

We wish to compute $P_{30}(t)$, the probability that the system enters E_0 (i.e., fails) on or before time t, given that it was in E_3 at $t = 0$. It is readily verified that the set of differential equations satisfied by $P_{33}(t)$, $P_{32}(t)$, $P_{31}(t)$ $P_{30}(t)$ are

$$
\begin{aligned}
P_{33}'(t) &= -\lambda_3 P_{33}(t) + \mu_2 P_{32}(t) \\
P_{32}'(t) &= \lambda_3 P_{33}(t) - (\lambda_2 + \mu_2) P_{32}(t) + \mu_1 P_{31}(t) \\
P_{31}'(t) &= \lambda_2 P_{32}(t) - (\lambda_1 + \mu_1) P_{31}(t) \\
P_{30}'(t) &= \lambda_1 P_{31}(t) \, .
\end{aligned}
\tag{7.32}
$$

Applying Laplace transforms to both sides of (7.32), we obtain the set of equations

$$
\begin{aligned}
(\lambda_3 + s) P_{33}^*(s) - \mu_2 P_{32}^*(s) &= 1 \\
-\lambda_3 P_{33}^*(s) + (\lambda_2 + \mu_2 + s) P_{32}^*(s) - \mu_1 P_{31}^*(s) &= 0 \\
-\lambda_2 P_{32}^*(s) + (\lambda_1 + \mu_1 + s) P_{31}^*(s) &= 0
\end{aligned}
\tag{7.33}
$$

and

$$
s P_{30}^*(s) = \lambda_1 P_{31}^*(s) \, .
$$

Solving for $P_{31}^*(s)$, we get

$$
P_{31}^*(s) = \frac{\begin{vmatrix} \lambda_3 + s & -\mu_2 & 1 \\ -\lambda_3 & \lambda_2 + \mu_2 + s & 0 \\ 0 & -\lambda_2 & 0 \end{vmatrix}}{\begin{vmatrix} \lambda_3 + s & -\mu_2 & 0 \\ -\lambda_3 & \lambda_2 + \mu_2 + s & -\mu_1 \\ 0 & -\lambda_2 & \lambda_1 + \mu_1 + s \end{vmatrix}} \, .
\tag{7.34}
$$

After reduction, we get

$$
P_{31}^*(s) = \frac{\lambda_2 \lambda_3}{s^3 + Bs^2 + Cs + D} = \frac{\lambda_2 \lambda_3}{(s - s_1)(s - s_2)(s - s_3)} \, ,
\tag{7.35}
$$

where the s_i are the characteristic roots and

$$
\begin{aligned}
B &= \lambda_1 + \lambda_2 + \lambda_3 + \mu_1 + \mu_2 \\
C &= \lambda_1 \lambda_2 + \lambda_1 \lambda_3 + \lambda_2 \lambda_3 + \mu_1 \lambda_3 + \mu_2 \lambda_1 + \mu_1 \mu_2 \\
D &= \lambda_1 \lambda_2 \lambda_3 \, .
\end{aligned}
\tag{7.36}
$$

Hence, expanding in partial fractions, $P_{31}^*(s)$ becomes

$$
P_{31}^*(s) = \lambda_2 \lambda_3 \left[\frac{A_{31}}{s - s_1} + \frac{A_{32}}{s - s_2} + \frac{A_{33}}{s - s_3} \right] ,
\tag{7.37}
$$

where

$$
\begin{aligned}
A_{31} &= \frac{1}{(s_1 - s_2)(s_1 - s_3)}, \\
A_{32} &= \frac{1}{(s_2 - s_1)(s_2 - s_3)}, \\
A_{33} &= \frac{1}{(s_3 - s_1)(s_3 - s_2)}.
\end{aligned}
\tag{7.38}
$$

Inverting (7.37), we obtain

$$
P_{31}(t) = \lambda_2 \lambda_3 \sum_{i=1}^{3} A_{3i} e^{s_i t}
\tag{7.39}
$$

and

$$
P_{30}'(t) = \lambda_1 \lambda_2 \lambda_3 \sum_{i=1}^{3} A_{3i} e^{s_i t}.
\tag{7.40}
$$

It should be noted that $P_{30}'(t)$ is the p.d.f. of the first passage time from $E_3 \to E_0$. Furthermore, $R_{30}(t)$, the probability that E_0 is not reached before time t, is given by

$$
R_{30}(t) = \int_{t}^{\infty} P_{30}'(\tau) d\tau = -\lambda_1 \lambda_2 \lambda_3 \sum_{i=1}^{3} \frac{A_{3i} e^{s_i t}}{s_i}.
\tag{7.41}
$$

7.2.2 Mean first passage times

A quantity of particular interest is ℓ_{30}, the mean first passage time from $E_3 \to E_0$. Integrating $R_{30}(t)$ from Equation (7.41) gives

$$
\ell_{30} = \lambda_1 \lambda_2 \lambda_3 \sum_{i=1}^{3} \frac{A_{3i}}{s_i^2}.
$$

But to get an explicit expression, one must compute the roots s_i. One can instead solve the system of linear equations

$$
\begin{aligned}
\ell_{30} &= \frac{1}{\lambda_3} + \ell_{20} \\
\ell_{20} &= \frac{1}{\lambda_2 + \mu_2} + \frac{\lambda_2}{\lambda_2 + \mu_2}\ell_{10} + \frac{\mu_2}{\lambda_2 + \mu_2}\ell_{30} \\
\ell_{10} &= \frac{1}{\lambda_1 + \mu_1} + \frac{\mu_1}{\lambda_1 + \mu_1}\ell_{20},
\end{aligned}
$$

the solution of which is given by

$$\ell_{30} = \frac{1}{\lambda_1} + \frac{1}{\lambda_3} + \frac{(\lambda_1 + \mu_1)(\lambda_3 + \mu_2)}{\lambda_1 \lambda_2 \lambda_3}$$

$$\ell_{20} = \frac{1}{\lambda_1} + \frac{(\lambda_1 + \mu_1)(\lambda_3 + \mu_2)}{\lambda_1 \lambda_2 \lambda_3}$$

$$\ell_{10} = \frac{1}{\lambda_1} + \frac{\mu_1}{\lambda_1 \lambda_2} + \frac{\mu_1 \mu_2}{\lambda_1 \lambda_2 \lambda_3}.$$

Still another method is the one based on the Laplace transform of $P'_{30}(t)$. The Laplace transform of $P'_{30}(t)$ is

$$L\{P'_{30}\}(s) = \lambda_1 P^*_{31}(s) = \frac{\lambda_1 \lambda_2 \lambda_3}{s^3 + Bs^2 + Cs + D} = \frac{D}{s^3 + Bs^2 + Cs + D}. \tag{7.42}$$

It is readily verified that

$$\ell_{30} = -\lambda_1 \left. \frac{dP^*_{31}(s)}{ds} \right|_{s=0}. \tag{7.43}$$

Differentiating the right-hand side of Equation (7.42) with respect to s and setting $s = 0$ gives

$$\ell_{30} = \frac{C}{D} = \frac{\lambda_1 \lambda_2 + \lambda_1 \lambda_3 + \lambda_2 \lambda_3 + \mu_1 \lambda_3 + \mu_2 \lambda_1 + \mu_1 \mu_2}{\lambda_1 \lambda_2 \lambda_3}$$

$$= \frac{1}{\lambda_1} + \frac{1}{\lambda_2} + \frac{1}{\lambda_3} + \frac{\mu_1}{\lambda_1 \lambda_2} + \frac{\mu_2}{\lambda_2 \lambda_3} + \frac{\mu_1 \mu_2}{\lambda_1 \lambda_2 \lambda_3} \tag{7.44}$$

$$= \frac{1}{\lambda_3} + \frac{1}{\lambda_2}\left(1 + \frac{\mu_2}{\lambda_3}\right) + \frac{1}{\lambda_1}\left(1 + \frac{\mu_1}{\lambda_2} + \frac{\mu_1 \mu_2}{\lambda_2 \lambda_3}\right).$$

Each of the three terms on the right-hand side of Equation (7.44) has a meaning:

$$\frac{1}{\lambda_3} = \ell_{32}, \quad \text{the mean first passage time from } E_3 \to E_2$$

$$\frac{1}{\lambda_2}\left(1 + \frac{\mu_2}{\lambda_3}\right) = \ell_{21}, \quad \text{the mean first passage time from } E_2 \to E_1$$

and

$$\frac{1}{\lambda_1}\left(1 + \frac{\mu_1}{\lambda_2} + \frac{\mu_1 \mu_2}{\lambda_2 \lambda_3}\right) = \ell_{10}, \quad \text{the mean first passage time from } E_1 \to E_0.$$

Similarly, for a four-unit system, one can verify that ℓ_{40}, the mean first passage time from $E_4 \rightarrow E_0$ is

$$
\begin{aligned}
\ell_{40} &= \ell_{43} + \ell_{32} + \ell_{21} + \ell_{10} \\
&= \frac{1}{\lambda_4} + \frac{1}{\lambda_3}\left(1 + \frac{\mu_3}{\lambda_4}\right) + \frac{1}{\lambda_2}\left(1 + \frac{\mu_2}{\lambda_3} + \frac{\mu_2\mu_3}{\lambda_3\lambda_4}\right) \qquad (7.45) \\
&\quad + \frac{1}{\lambda_1}\left(1 + \frac{\mu_1}{\lambda_2} + \frac{\mu_1\mu_2}{\lambda_2\lambda_3} + \frac{\mu_1\mu_2\mu_3}{\lambda_2\lambda_3\lambda_4}\right).
\end{aligned}
$$

More generally, for an n-unit system, it can be shown that ℓ_{n0}, the mean first passage time from $E_n \rightarrow E_0$, is

$$
\begin{aligned}
\ell_{n0} &= \ell_{n,n-1} + \ell_{n-1,n-2} + \cdots + \ell_{10} \qquad\qquad (7.46) \\
&= \frac{1}{\lambda_n} + \frac{1}{\lambda_{n-1}}\left(1 + \frac{\mu_{n-1}}{\lambda_n}\right) + \frac{1}{\lambda_{n-2}}\left(1 + \frac{\mu_{n-2}}{\lambda_{n-1}} + \frac{\mu_{n-2}\mu_{n-1}}{\lambda_{n-1}\lambda_n}\right) \\
&\quad + \cdots + \frac{1}{\lambda_1}\left(1 + \frac{\mu_1}{\lambda_2} + \frac{\mu_1\mu_2}{\lambda_2\lambda_3} + \cdots + \frac{\mu_1\mu_2\cdots\mu_{n-1}}{\lambda_2\lambda_3\cdots\lambda_n}\right).
\end{aligned}
$$

For another approach to first passage time problems see Barlow and Proschan [3].

7.2.3 Other initial states

Returning to the three-unit case, we note that all computations were based on the assumption that the system was in E_3 at time $t = 0$. Let us now see what happens if the initial state is E_2 or E_1. If the initial state is E_2, we would get precisely the same set of differential Equations (7.32) except that $P_{3j}(t)$ and $P'_{3j}(t)$ would be replaced by $P_{2j}(t)$ and $P'_{2j}(t)$. Instead of the set of Equations (7.33) we would get the set

$$
\begin{aligned}
(\lambda_3 + s)P_{23}^*(s) - \mu_2 P_{22}^*(s) &= 0 \\
-\lambda_3 P_{23}^*(s) + (\lambda_2 + \mu_2 + s)P_{22}^*(s) - \mu_1 P_{21}^*(s) &= 1 \qquad (7.47) \\
-\lambda_2 P_{22}^*(s) + (\lambda_1 + \mu_1 + s)P_{21}^*(s) &= 0
\end{aligned}
$$

and

$$
sP_{20}^*(s) = \lambda_1 P_{21}^*(s).
$$

It is readily verified that

$$
P_{21}^*(s) = \frac{\lambda_2(\lambda_3 + s)}{(s - s_1)(s - s_2)(s - s_3)}, \qquad (7.48)
$$

$$
P'_{20}(t) = \lambda_1\lambda_2\lambda_3 \sum_{i=1}^{3} A_{2i} e^{s_i t}. \qquad (7.49)
$$

and

$$R_{20}(t) = -\lambda_1\lambda_2\lambda_3 \sum_{i=1}^{3} \frac{A_{2i}e^{s_i t}}{s_i} \qquad (7.50)$$

where

$$A_{2i} = \frac{\lambda_3 + s_i}{\lambda_3} A_{3i}, \qquad i = 1, 2, 3.$$

If the initial state is E_1, it is readily verified that

$$P_{11}^*(s) = \frac{(\lambda_2 + s)(\lambda_3 + s) + \mu_2 s}{(s - s_1)(s - s_2)(s - s_3)}, \qquad (7.51)$$

$$P_{10}'(t) = \lambda_1\lambda_2\lambda_3 \sum_{i=1}^{3} A_{1i}e^{s_i t} \qquad (7.52)$$

and

$$R_{10}(t) = -\lambda_1\lambda_2\lambda_3 \sum_{i=1}^{3} \frac{A_{1i}e^{s_i t}}{s_i}, \qquad i = 1, 2, 3 \qquad (7.53)$$

where

$$A_{1i} = \frac{(\lambda_2 + s_i)(\lambda_3 + s_i) + \mu_2 s_i}{\lambda_2\lambda_3} A_{3i}, \qquad i = 1, 2, 3.$$

Remark 1. It is useful to note that $\ell_{30} = -\left(\frac{1}{s_1} + \frac{1}{s_2} + \frac{1}{s_3}\right)$. This is proved as follows:

According to (7.44), $\ell_{30} = C/D$. But, from the equation $s^3 + Bs^2 + Cs + D = (s - s_1)(s - s_2)(s - s_3)$, we conclude that $C = s_1 s_2 + s_1 s_3 + s_2 s_3$ and $D = -s_1 s_2 s_3$. Algebraic simplification then yields the desired result.

More generally, it can be shown, for the n-unit repairable system, that ℓ_{n0} is given by

$$\ell_{n0} = -\sum_{i=1}^{n} \frac{1}{s_i}, \qquad (7.54)$$

where the s_i are the roots of a polynomial of degree n and all of the $s_i < 0$. The key fact needed to prove Equation (7.54) is that $A(s) = L\{P_{n0}'\}(s)$, the Laplace transform of $P_{n0}'(t)$ (the p.d.f. of the first passage time from $E_n \rightarrow E_0$) has the form

$$A(s) = \prod_{i=1}^{n} \frac{\lambda_i}{s - s_i} = \prod_{i=1}^{n} \frac{|s_i|}{s - s_i}. \qquad (7.55)$$

Hence,

$$\log A(s) = K - \sum_{i=1}^{n} \log(s - s_i), \qquad (7.56)$$

where K is a constant. Differentiating both sides of Equation (7.56) with respect to s gives

$$\frac{A'(s)}{A(s)} = -\sum_{i=1}^{n} \frac{1}{s - s_i}. \tag{7.57}$$

Hence, $A'(s)]_{s=0} = \sum_{i=1}^{n}(s_i)^{-1}$, since $A(0) = 1$. But

$$\ell_{n0} = -A'(s)]_{s=0} = -\sum_{i=1}^{n} \frac{1}{s_i}$$

and this establishes Equation (7.54).

Remark 2. Formulae (7.40) and (7.41) give the exact solutions for $P'_{30}(t)$ and $R_{30}(t)$. It should be noted that these formulae can be expressed completely in terms of the roots s_1, s_2, s_3 since $D = \lambda_1\lambda_2\lambda_3 = -s_1 s_2 s_3$. For the n-unit repairable system, it follows by inverting Equation (7.55) that

$$P'_{n0}(t) = \prod_{i=1}^{n} \lambda_i \sum_{i=1}^{n} A_i e^{s_i t} = \prod_{i=1}^{n} |s_i| \sum_{i=1}^{n} A_i e^{s_i t}, \tag{7.58}$$

where

$$A_i = \prod_{j \neq i}(s_i - s_j)^{-1}. \tag{7.59}$$

Furthermore, $R_{n0}(t)$ is given by

$$R_{n0}(t) = \int_{t}^{\infty} P'_{n0}(\tau)d\tau = \prod_{i=1}^{n} |s_i| \sum_{i=1}^{n} \frac{A_i e^{s_i t}}{|s_i|}. \tag{7.60}$$

Suppose that the absolute value of the roots are arranged in increasing order of size

$$|s_{(1)}| < |s_{(2)}| < \cdots < |s_{(n)}|, \tag{7.61}$$

where $|s_{(j)}|$, $j = 1, 2, \ldots, n$ is the jth smallest member of the set $\{|s_1|, |s_2|, \ldots, |s_n|\}$. Suppose further that

$$|s_{(1)}| \ll |s_{(j)}|, \qquad j = 2, \ldots, n \tag{7.62}$$

where $\ll$ means "much smaller than." Then it can be verified that $e^{s_{(1)}t}$ is a good approximation for $R_{n0}(t)$. In many practical problems condition (7.62) is satisfied. This will occur when the μ_i are considerably larger than the corresponding λ_i.

If condition (7.62) holds, then $s_{(1)}$ can be approximated very easily. We know, on the one hand, from Equation (7.46) that ℓ_{n0} can be expressed completely in terms of $\{\lambda_1, \lambda_2, \ldots, \lambda_n\}$, $\{\mu_0, \mu_1, \ldots, \mu_{n-1}\}$ and, on the other hand, from Equation (7.54) that $\ell_{n0} = -\sum_{i=1}^{n} s_{(i)}^{-1}$. But, if condition (7.62) holds, then

$s_{(1)} \cong -1/\ell_{n0}$ and hence

$$R_{n0}(t) \cong e^{-t/\ell_{n0}} \cong e^{s_{(1)}t}, \tag{7.63}$$

where $\cong$ means "is well approximated by." One can also use Equation (7.60) to establish the approximation. Assume, without loss of generality, that $s_{(1)} = s_1$, then only the first term in the sum of Equation (7.60) is important. Moreover, $(\prod |s_i|)A_1/|s_1| \cong 1$, implying the approximation (7.63).

The fact that approximation $e^{s_{(1)}t}$ is a good approximation for the first passage time distribution from $E_n \to E_0$ under fairly modest conditions, often satisfied in practice, is discussed in some detail in Chapter 6 of Gnedenko et al. [30].

7.2.4 Examples

Example 1. Consider a three-unit repairable system with one repair facility. Suppose that all units are active. Assume that $\lambda = .1/\text{hour}$ and $\mu = 1/\text{hour}$. Find $R_{30}(t)$.

Solution: In this problem $\lambda_3 = 3\lambda$, $\lambda_2 = 2\lambda$, $\lambda_1 = \lambda$; $\mu_2 = \mu_1 = \mu$. The equation $s^3 + Bs^2 + Cs + D = 0$ becomes $s^3 + 2.6s^2 + 1.51s + .006 = 0$. The exact solution for $R_{30}(t)$ involves finding the roots s_1, s_2 and s_3 and substituting in Equation (7.41). To five significant figures, the roots are: $s_{(1)} = -.00400$, $s_{(2)} = -.86766$ and $s_{(3)} = -1.72833$. Note that indeed $|s_{(1)}| \ll |s_{(2)}|, |s_{(3)}|$. The approximation (7.63) is obtained by computing

$$\ell_{30} = \frac{C}{D} = \frac{1.51}{.006} = 251.67.$$

Thus $R_{30}(t) \cong e^{-t/251.67} = e^{-.0039735t}$.

Note that the approximate value of $s_{(1)}$ obtained from the equation $s_{(1)} \cong -1/\ell_{30} = -.0039735$ is in good agreement with the exact value of $s_{(1)}$ (namely, relative error of $-.66\%$).

Example 2. Let us assume that there are four units, each active and having failure rate λ, and two repair facilities, each having repair rate μ. We wish to find the distribution of the first time that a unit will be waiting for service. If we let $E_i =$ state in which there are i good units, then the problem involves finding the first passage time distribution from E_4 to E_1. It can be verified that the p.d.f. of the first passage time from $E_4 \to E_1$ is given by

$$P'_{41}(t) = \lambda_2\lambda_3\lambda_4[A_1 e^{s_1 t} + A_2 e^{s_2 t} + A_e e^{s_3 t}],$$

where s_1, s_2, s_3 are the roots of the equation $s^3 + Bs^2 + Cs + D = 0$ with

$$
\begin{aligned}
B &= \lambda_2 + \lambda_3 + \lambda_4 + \mu_2 + \mu_3, \\
C &= \lambda_2\lambda_3 + \lambda_2\lambda_4 + \lambda_3\lambda_4 + \mu_2\lambda_4 + \mu_3\lambda_2 + \mu_2\mu_3, \\
D &= \lambda_2\lambda_3\lambda_4.
\end{aligned}
$$

But $\lambda_4 = 4\lambda$, $\lambda_3 = 3\lambda$, $\lambda_2 = 2\lambda$, $\mu_3 = \mu$ and $\mu_2 = 2\mu$. It follows that $B = 9\lambda + 3\mu$, $\quad C = 26\lambda^2 + 10\lambda\mu + 2\mu^2$ and $D = 24\lambda^3$. Moreover,

$$A_1 = \frac{1}{(s_1 - s_2)(s_1 - s_3)},$$

$$A_2 = \frac{1}{(s_2 - s_1)(s_2 - s_3)},$$

$$A_3 = \frac{1}{(s_3 - s_1)(s_3 - s_2)}.$$

Thus, ℓ_{41}, the mean first passage time from $E_4 \to E_1$ is given by

$$\ell_{41} = \frac{C}{D} = \frac{26\lambda^2 + 10\lambda\mu + 2\mu^2}{24\lambda^3} = \frac{1}{12\lambda}(13 + 5x + x^2),$$

where $x = \mu/\lambda$. If $\lambda = .1$, $\mu = 1$, the equation $s^3 + Bs^2 + Cs + D = 0$ becomes $s^3 + 3.9s^3 + 3.26s + .024 = 0$. To five significant figures, the roots are $s_{(1)} = -.0074278$, $s_{(2)} = -1.20000$, $s_{(3)} = -2.69257$. Note that indeed $|s_{(1)}| \ll |s_{(2)}|, |s_{(3)}|$. Hence, $R_{41}(t) \cong e^{-t/\ell_{41}}$, where

$$\ell_{41} = \frac{C}{D} = \frac{3.26}{.024} = 135.833,$$

i.e., $R_{41}(t) = e^{-.007362t}$. Note that the approximate value of $s_{(1)}$ obtained from the equation $s_{(1)} \cong -\frac{1}{\ell_{41}} = -.007362$ is in good agreement with the exact value of $s_{(1)}$ (relative error of $-.89\%$).

Example 3 (see Epstein [22]). It is useful to point out that for the n-machine repair problem with transition probabilities given by Table 6.3, long-run system availability (and unavailability) can be expressed simply in terms of appropriate mean first passage times. Consider, for example, the n-unit repairable system for which $\{E_1, E_2, \ldots, E_n\}$ are good states and E_0 is a bad state. Since long-run system availability is independent of the starting state, let us assume that the system starts in E_0. Of particular interest are cycles whose average length is composed of two parts:

(i) $\ell_{01} = 1/\mu_0$ – the average length of time spent in E_0 before entering E_1 (average system repair or down-time in a single cycle).

(ii) ℓ_{10} – the average length of time to return to E_0 from E_1 (average system up-time in a single cycle).

It follows from the (strong) law of large numbers that the unavailability, U, is given by

$$U = \frac{1/\mu_0}{\ell_{10} + 1/\mu_0} = \frac{1}{\mu_0\ell_{10} + 1}. \qquad (7.64)$$

An alternative proof of Equation (7.64) is obtained by noting that according to Equation (7.46)

$$\ell_{10} = \frac{1}{\lambda_1} + \frac{\mu_1}{\lambda_1\lambda_2} + \frac{\mu_1\mu_2}{\lambda_1\lambda_2\lambda_3} + \cdots + \frac{\mu_1\mu_2\cdots\mu_{n-1}}{\lambda_1\lambda_2\ldots\lambda_n} . \tag{7.65}$$

Multiplying both sides by μ_0 and using Equation (6.21), we obtain

$$\mu_0\ell_{10} = \frac{1-p_0}{p_0} = \frac{1-U}{U} . \tag{7.66}$$

Solving Equation (7.66) for U gives Equation (7.64).

In the preceding discussion, we have assumed that $\{E_1, E_2, \ldots, E_n\}$ are good states and that E_0 is a bad state for the n-unit system. More generally, the states $\{E_j\}$, $j = 0, 1, 2, \ldots, n$ can be divided into two disjoint sets A and $\bar{A}$ such that set A consists of all those states E_j with subscripts j such that $r \leq j \leq n$ and set $\bar{A}$ consists of all those states E_j with subscripts j such that $0 \leq j \leq r - 1$. The states belonging to $A(\bar{A})$ are considered to be good (bad) states for the system. Since long-run system availability is independent of the starting state, let us assume that the system is initially in E_{r-1} (the "first" bad state that can be reached from a good state). Of particular interest are cycles whose average length is composed of two parts:

(i) $\ell_{r-1,r}$ – the mean time to pass from E_{r-1} into E_r (the "first" good state that can be reached from a bad state). It is also the average system repair of down-time in a single cycle.

(ii) $\ell_{r,r-1}$ – the mean time to return to E_{r-1} from E_r (the average system up-time in a single cycle).

It follows from the (strong) law of large numbers that the long-run system availability, A, is given by

$$A = \frac{\ell_{r,r-1}}{\ell_{r-1,r} + \ell_{r,r-1}} . \tag{7.67}$$

To show directly that the right-hand side of Equation (7.67) equals $\sum_{j=r}^{n} p_j$ (the long-run system availability by definition), we proceed as follows: According to Equation (7.46),

$$\ell_{r,r-1} = \frac{1}{\lambda_r} \left(1 + \frac{\mu_r}{\lambda_{r+1}} + \frac{\mu_r\mu_{r+1}}{\lambda_{r+1}\lambda_{r+2}} + \cdots + \frac{\mu_r\mu_{r+1}\cdots\mu_{n-1}}{\lambda_{r+1}\lambda_{r+2}\ldots\lambda_n} \right) . \tag{7.68}$$

Multiplying both sides by $\mu_0\mu_1 \ldots \mu_{r-1}/[\lambda_1\lambda_2 \ldots \lambda_{r-1}]$ (for $r = 1$, we multiply by μ_0). This was done in going from Equation (7.65) to Equation (7.66). For $r \geq 2$, utilizing Equation (6.21) yields

$$\ell_{r,r-1} = \frac{\lambda_1\lambda_2\ldots\lambda_{r-1}}{\mu_0\mu_1\ldots\mu_{r-1}} \cdot \frac{\sum_{j=r}^{n} p_j}{p_0} . \tag{7.69}$$

It can be verified that $\ell_{r-1,r}$, the mean (first passage) time from E_{r-1} into E_r, is given by

$$\ell_{r-1,r} = \frac{1}{\mu_{r-1}} \left[1 + \frac{\lambda_{r-1}}{\mu_{r-2}} + \frac{\lambda_{r-1}\lambda_{r-2}}{\mu_{r-2}\mu_{r-3}} + \cdots + \frac{\lambda_{r-1}\lambda_{r-2}\ldots\lambda_2\lambda_1}{\mu_{r-2}\mu_{r-3}\ldots\mu_1\mu_0} \right].$$
(7.70)

For $r = 1$, $\ell_{r-1,r} = \ell_{0,1} = 1/\mu_0$. For $r \geq 2$, we multiply both sides of Equation (7.70) by $\mu_0\mu_1 \ldots \mu_{r-1}/[\lambda_1\lambda_2 \ldots \lambda_{r-1}]$. Utilizing Equation (6.21), this yields

$$\ell_{r-1,r} = \frac{\lambda_1\lambda_2\ldots\lambda_{r-1}}{\mu_0\mu_1\ldots\mu_{r-1}} \cdot \frac{\sum_{j=0}^{r-1} p_j}{p_0}.$$
(7.71)

Hence, $\ell_{r,r-1}/(\ell_{r-1,r} + \ell_{r,r-1}) = \sum_{j=r}^{n} p_j$ and Equation (7.67) is verified algebraically since $A = \sum_{j=r}^{n} p_j$.

Example 4. Suppose that we have an n-unit repairable system and ask the following questions: If we look at the system at some randomly chosen time far in the future, where the system is in a good state. What is the expected length of time before the system enters E_0 (the failed state)? What is the probability that the system will not fail for the next τ units of time? It is readily verified that the answer to the first question is $\sum_{j=1}^{n} p_j \ell_{j0}$, where the p_j satisfy Equation (6.21) and the ℓ_{j0} can be determined from Equation (7.46). It also can be verified that the answer to the second question is given by $\sum_{j=1}^{n} p_j R_{j0}(\tau)$. The $R_{j0}(\tau)$ can be determined by the methods given in this chapter.

7.3 Repair time follows a general distribution

7.3.1 First passage time

In our treatment of the reliability of n-unit repairable systems, we have assumed that the time to failure and time to repair a unit are exponentially distributed. These assumptions make the analysis easy since the state of the system at any time t is described completely by $N(t)$, the number of units which are in a good state at time t. Such a simple description is possible because of the *nonaging* * property of the exponential distribution. Another consequence of the assumption of exponentially distributed unit time to failure and repair distributions is that $N(t)$ is a Markov process for all $t \geq 0$. By this, we mean that if we are given the entire history of $N(t)$ up to t^*, then for all $t > t^*$, the

---

* Consider a device whose length of life is distributed with p.d.f. $f(t) = \lambda e^{-\lambda t}$, $t \geq 0$. Then $P(T > t + \tau \mid T > t) = e^{-\lambda\tau}$, i.e., given that the device has survived up to time t, the probability that it survives an additional amount of time τ is exactly the same as the probability that a new device placed in service at $t = 0$ survives for a length of time τ. This property is often referred to as the *nonaging* property of the exponential distribution.

conditional distribution of $N(t)$ given this history is identical with the conditional distribution of $N(t)$ given $N(t^*)$ alone. Put another way, the "future" (i.e., $t > t^*$) stochastic behavior of the system depends not on the entire past $[0, t^*]$, but only upon the immediate past, namely the state of the system at t^*. The fact that $N(t)$ is a Markov process is of fundamental importance and it is this fact that makes the reliability analysis straightforward. However, $N(t)$ is no longer a Markov process if either unit time to failure or unit time to repair distributions or both are not exponential. Additional random variables are needed to complete the description of the state of the system at time t and this is one way of making a process which is non-Markovian in time, Markovian in a space of higher dimensions. For more details on the use of supplementary variables to make a non-Markovian process Markovian, see Cox and Miller [10]. For example, consider the simple case of a one-unit repairable system with general unit time to failure c.d.f. $F(x)$ and general unit time to repair c.d.f. $G(y)$ and let $N(t) = 1(0)$, if the system is up (down). Here $N(t)$ is no longer a Markov process over time. To make the process Markovian, we must specify at each time t, not only $N(t)$, but also the length of time $X_t(Y_t)$ that the unit has been operating without failure since its most recent entrance into the up state, if $N(t) = 1$ (that the unit has been under repair since its most recent entrance into the failed state, if $N(t) = 0$). The two-dimensional stochastic process $(N(t), X_t(Y_t))$ is Markovian. By this, we mean that if we are given the entire history of $(N(t), X_t(Y_t))$ up to time t^*, then for all $t > t^*$, the conditional distribution of $(N(t), X_t(Y_t))$ given this history is identical with the conditional distribution of $(N(t), X_t(Y_t))$ given $(N(t^*), X_{t^*}(Y_{t^*}))$ alone.

In [28], Gaver analyzes the stochastic behavior over time of a two-unit repairable system with a single repairman, where the time to failure of a unit is exponentially distributed and where the time to repair a unit is a random variable Y with c.d.f. $G(y)$, p.d.f. $g(y) = G'(y)$, and associated repair rate $h(y) = g(y)/[1 - G(y)]$ (recall, $h(y)dy$ is the probability that a device will be repaired in the time interval $(y, y + dy)$ given that it has not been repaired by time y). It is particularly instructive to solve the first passage time problem for such a system, because it shows clearly the complications introduced by nonexponentiality of the repair distribution. It is also a good example of how one can make a stochastic process, which is non-Markovian in time, Markovian by adding a supplementary variable to describe the state of the system. At any instant t, the system is either in state 2 (both units are good); $(1, x)$ (1 unit is good and 1 unit has failed and has been under repair for a length of time x); or $(0, x)$ (both units have failed; one unit is waiting to be serviced and the other unit has been under repair for a length of time x). System failure is said to occur as soon as state $(0, x)$ is reached for any $x \geq 0$. The transitions from state to state in the time interval $(t, t + \Delta]$ are given in Table 7.1.

Let $P_2(t)$ be the probability of being in state 2 at time t; let $P_1(x, t)$ be a

Table 7.1

Transition	Probability of transition in $(t, t + \Delta)$
$2 \longrightarrow 2$	$1 - \lambda_2\Delta + o(\Delta)$
$2 \longrightarrow (1, x = 0)$	$\lambda_2\Delta + o(\Delta)$
$2 \longrightarrow 0$	$o(\Delta)$
$(1, x) \longrightarrow 2$	$h(x)\Delta + o(\Delta)$
$(1, x) \longrightarrow (1, x + \Delta)$	$1 - (\lambda_1 + h(x))\Delta + o(\Delta)$
$(1, x) \longrightarrow (0, x + \Delta)$	$\lambda_1\Delta + o(\Delta)$
$0 \longrightarrow 0$	1

density (*not* a density function) in the sense that $P_1(x, t)dx$ is the probability
that the system occupies states lying between $(1, x)$ and $(1, x + dx)$ at time t;
and let $P_0(t)$ be the probability that the system fails on or before time t. Then
$P_0(t)$ is the c.d.f. of the first passage time into the failed state and $P_0'(t)$ is the
p.d.f. of the first passage time. Using the transition probabilities in Table 7.1,
we obtain the following system of difference equations relating the probability
that the system is in certain states at time $t + \Delta$ with the probability of being
in various states at time t:

$$P_2(t + \Delta) = P_2(t)[1 - \lambda_2\Delta] + \left[\int_0^\infty P_1(x, t)h(x)dx \right] \Delta + o(\Delta),$$

$$P_1(x + \Delta, t + \Delta) = P_1(x, t)[1 - (h(x) + \lambda_1)\Delta] + o(\Delta),$$

$$\text{(7.72)}$$

$$P_0(t + \Delta) = \left[\lambda_1 \int_0^\infty P_1(x, t)dx \right] \Delta + P_0(t) + o(\Delta).$$

Letting $\Delta \to 0$, we obtain the set of integro-differential equations:

$$P_2'(t) = -\lambda_2 P_2(t) + \int_0^\infty P_1(x, t)h(x)dx,$$

$$\frac{\partial P_1(x, t)}{\partial x} = \frac{\partial P_1(x, t)}{\partial t} = -[\lambda_1 + h(x)]P_1(x, t), \qquad \text{(7.73)}$$

and

$$P_0'(t) = \lambda_1 \int_0^\infty P_1(x, t)dx.$$

There are two boundary conditions associated with the system of Equations
(7.73). One of them is obtained by noting from Table 7.1 that

$$P_1(0, t)\Delta = \lambda_2 P_2(t)\Delta + o(\Delta). \qquad \text{(7.74)}$$

Letting $\Delta \to 0$, it follows that

$$P_1(0, t) = \lambda_2 P_2(t). \qquad \text{(7.75)}$$

Assuming that both units are initially in a good state gives the second boundary condition

$$P_2(0) = 1, \qquad P_1(0) = 0, \qquad P_0(0) = 0. \tag{7.76}$$

As before, we use Laplace transform method to solve the system of equations (7.73) subject to the boundary conditions (7.75) and (7.76). To this end, we define the Laplace transforms

$$P_2^*(s) = \int_0^\infty e^{-st} P_2(t)dt \tag{7.77}$$

and

$$P_1^*(x, s) = \int_0^\infty e^{-st} P_1(x, t)dt.$$

Applying Laplace transforms to the first two equations of (7.73) and to (7.75) and imposing the boundary conditions (7.76) yields the set of equations

$$sP_2^*(s) - 1 = -\lambda_2 P_2^*(s) + \int_0^\infty P_1^*(x, s)h(x)dx$$

$$sP_1^*(x, s) + \frac{\partial}{\partial x}P_1^*(x, s) = -[\lambda_1 + H(x)]P_1^*(x, s)$$

$$\lambda_2 P_2^*(s) = P_1^*(0, s). \tag{7.78}$$

Solving the second of these equations gives

$$P_1^*(x, s) = C(s)e^{-(s+\lambda_1)x - H(x)} \tag{7.79}$$

where $H(x) = \int_0^x h(y)dy$.

To get $C(s)$ we set $x = 0$ and use the third equation in (7.78). This gives

$$C(s) = P_1^*(0, s) = \lambda_2 P_2^*(s). \tag{7.80}$$

Substituting (7.79) into the first equation of (7.78) gives

$$sP_2^*(s) - 1 = -\lambda_2 P_2^*(s) + \lambda_2 P_2^*(s) \int_0^\infty e^{-(s+\lambda_1)x}h(x)e^{-H(x)}dx. \tag{7.81}$$

But $h(x)e^{-H(x)} = g(x)$, the p.d.f. of the time to repair distribution. We define $g^*(s)$, the Laplace transform of the repair time p.d.f., as

$$g^*(s) = \int_0^\infty e^{-sy}g(y)dy. \tag{7.82}$$

Consequently,

$$\int_0^\infty e^{-(s+\lambda_1)x}g(x)dx = g^*(s + \lambda_1) \tag{7.83}$$

and so

$$P_2^*(s) = \frac{1}{s + \lambda_2[1 - g^*(s + \lambda_1)]}. \tag{7.84}$$

Substituting into Equation (7.79) gives

$$
\begin{aligned}
P_1^*(x, s) &= \frac{\lambda_2}{s + \lambda_2[1 - g^*(s + \lambda_1)]} e^{-(s+\lambda_1)x} e^{-H(x)} \\
&= \frac{\lambda_2}{s + \lambda_2[1 - g^*(s + \lambda_1)]} e^{-(s+\lambda_1)x} [1 - G(x)] . \quad (7.85)
\end{aligned}
$$

Let $p_T^*(s)$ be the Laplace transform of $P_0'(t)$ then, using the last equation in (7.73), one gets

$$
\begin{aligned}
p_T^*(s) &= L\{P_0'\}(s) = E\left(e^{-sT}\right) = \int_0^\infty e^{-st} P_0'(t) dt \\
&= \lambda_1 \int_0^\infty P_1^*(x, s) dx \\
&= \frac{\lambda_1 \lambda_2 [1 - g^*(s + \lambda_1)]}{\{s + \lambda_2[1 - g^*(s + \lambda_1)]\}(s + \lambda_1)} . \quad (7.86)
\end{aligned}
$$

It is possible, at least in principle, to invert the right-hand side of Equation (7.86) and to then find an explicit formula for the p.d.f. of the first time to system failure T. The mean time to first system failure is given by

$$
MTFSF = -\frac{d}{ds} [p_T^*(s)]_{s=0} = \frac{\lambda_1 + \lambda_2 - \lambda_2 g^*(\lambda_1)}{\lambda_1 \lambda_2 [1 - g^*(\lambda_1)]} . \quad (7.87)
$$

As a matter of technical convenience, we have assumed the existence of $g(y) = G'(y)$ in establishing Equations (7.86) and (7.87). These formulae are, however, true even if $G'(y)$ does not exist. If this is the case, $g^*(s)$ is defined as $\int_0^\infty e^{-sy} dG(y)$, the Laplace-Stieltjes transform of $G(y)$.

7.3.2 Examples

Example 1. Suppose that we have a two-unit repairable system, both units active with unit time to failure p.d.f. $f(t) = \lambda e^{-\lambda t}$, $t \geq 0$ and a single repairman with time to repair p.d.f. $g(y) = \mu e^{-\mu y}$, $y \geq 0$. This is the case treated in Section 7.1.1. Clearly $\lambda_2 = 2\lambda$ and $\lambda_1 = \lambda$. Also $g^*(s) = \int_0^\infty e^{-sy} \mu e^{-\mu y} dy = \mu/(s + \mu)$. Substituting into Equation (7.86) gives

$$
p_T^*(s) = \frac{2\lambda^2}{s + 2\lambda \left[1 - \frac{\mu}{s+\lambda+\mu}\right]} \cdot \frac{1 - \frac{\mu}{s+\lambda+\mu}}{s + \lambda} = \frac{2\lambda^2}{s^2 + (3\lambda + \mu)s + 2\lambda^2} ,
$$

which agrees with Equation (7.5). The determination of $P_0'(t)$ from $p_T^*(s)$ is given in Section 7.1. Also

$$
E(T) = MTFSF = \frac{3\lambda - 2\lambda\frac{\mu}{\lambda+\mu}}{2\lambda^2 \left[1 - \frac{\mu}{\lambda+\mu}\right]} = \frac{3\lambda + \mu}{2\lambda^2} ,
$$

which agrees with Equation (7.12).

Example 2. Consider a two-unit system, both units active with unit time to failure p.d.f. $f(t) = \lambda e^{-\lambda t}$, $t \geq 0$, and a single repairman, who takes a constant amount of time A to make a repair. Hence the c.d.f. of the time to repair is

$$
\begin{aligned}
G(y) &= 0, & y < A \\
&= 1, & y \geq A.
\end{aligned}
$$

The Laplace-Stieltjes transform of $G(y)$ is

$$
g^*(s) = \int_0^\infty e^{-sy} dG(y) = e^{-As},
$$

and

$$
p_T^*(s) = \frac{2\lambda^2 \left[1 - e^{-A(s+\lambda)}\right]}{\left\{s + 2\lambda \left[1 - e^{-A(s+\lambda)}\right]\right\}(s+\lambda)}.
$$

By Equation (7.87), the expected waiting time to the first system failure is given

$$
E(T) = \frac{3 - 2e^{-A\lambda}}{2\lambda \left[1 - e^{-A\lambda}\right]}.
$$

7.3.3 Steady-state probabilities

Up to this point, we have made the failed state an absorbing state. The reason for this is that we were interested in finding the first passage time distribution into the failed state. Of course, in real life repair would be performed when the two-unit system has failed, and system operation would resume as soon as one of the failed units has been repaired. What we would like to do now is find the steady-state (long-run) probability that two units are operating, one unit is operating, and no units are operating.

Let $P_2(t)$ be the probability that both units are in a good state at time t; let $P_1(x,t)dx$ be the probability that at time t there is one good unit and one failed unit and that the failed unit has been under repair for a length of time lying between x and $x + dx$; and let $P_0(x,t)dx$ be the probability that at time t both units have failed and that one of the failed units has been under repair for a length of time lying between x and $x + dx$ (repair has not been started on the other failed unit because of the assumption that there is only one repairman).

It is readily verified that $P_2(t)$, $P_1(x,t)$, and $P_0(x,t)$ satisfy the system of difference equations

$$
\begin{aligned}
P_2(t+\Delta) &= (1 - \lambda_2\Delta)P_2(t) + \Delta \int_0^\infty P_1(x,t)h(x)dx + o(\Delta) \\
P_1(x+\Delta,t+\Delta) &= P_1(x,t)[1 - (h(x) + \lambda_1)\Delta] + o(\Delta) \qquad (7.88) \\
P_0(x+\Delta,t+\Delta) &= P_0(x,t)[1 - h(x)\Delta] + \lambda_1 P_1(x,t)\Delta + o(\Delta),
\end{aligned}
$$

subject to the boundary conditions

$$
\begin{aligned}
P_1(0,t)\Delta &= \lambda_2 P_2(t)\Delta + \Delta \int_0^\infty P_0(x,t)h(x)dx + o(\Delta) \\
P_0(0,t) &\equiv 0 \quad \text{and} \quad P_2(0) = 1 .
\end{aligned}
\tag{7.89}
$$

Letting $\Delta \to 0$, Equations (7.88) and (7.89) yield

$$
P_2'(t) = -\lambda_2 P_2(t) + \int_0^\infty P_1(x,t)h(x)dx
$$

$$
\frac{\partial P_1(x,t)}{\partial x} + \frac{\partial P_1(x,t)}{\partial t} = -[h(x) + \lambda_1]P_1(x,t)
\tag{7.90}
$$

$$
\frac{\partial P_0(x,t)}{\partial x} + \frac{\partial P_0(x,t)}{\partial t} = -h(x)P_0(x,t) + \lambda_1 P_1(x,t) ,
$$

subject to the boundary conditions

$$
\begin{aligned}
P_1(0,t) &= \lambda_2 P_2(t) + \int_0^\infty P_0(x,t)h(x)dx \\
P_0(0,t) &\equiv 0 \quad \text{and} \quad P_2(0) = 1 .
\end{aligned}
\tag{7.91}
$$

We are particularly interested in finding the steady-state values $p_2 = \lim_{t\to\infty} P_2(t)$, $p_1(x) = \lim_{t\to\infty} P_1(x,t)$, and $p_0(x) = \lim_{t\to\infty} P_0(x,t)$, which can be shown to exist. The limit p_2 is the steady-state probability of finding both units in a good state. The limit $p_1(x)dx$ is the steady-state probability of finding one unit in a good state, one unit in a failed state, and with the repair time on the failed unit lying between x and $x + dx$. Similarly, $p_0(x)dx$ is the steady-state probability of finding both units down and with the repair time on the failed unit which is being repaired, lying between x and $x + dx$. Similarly, $p_1 = \int_0^\infty p_1(x)dx$ is the long-run probability of finding one good unit and one unit under repair and $p_0 = \int_0^\infty p_0(x)dx$ is the long-run probability that both units are down.

Letting $t \to \infty$, it is readily verified from Equation (7.90) that p_2, $p_1(x)$, and $p_0(x)$ satisfy

$$
\begin{aligned}
\lambda_2 p_2 &= \int_0^\infty p_1(x)h(x)dx \\
p_1'(x) &= -[\lambda_1 + h(x)]p_1(x) \\
p_0'(x) &= -h(x)p_0(x) + \lambda_1 p_1(x) ,
\end{aligned}
\tag{7.92}
$$

subject to the boundary conditions

$$
p_1(0) = \lambda_2 p_2 + \int_0^\infty p_0(x)h(x)dx \quad \text{and} \quad p_0(0) = 0 .
$$

Solving Equations (7.92) we obtain

$$
p_1(x) = Ae^{-\lambda_1 x + H(x)}
$$

and

$$p_0(x) = e^{-H(x)} \left\{ C - Ae^{-\lambda_1 x} \right\} . \tag{7.93}$$

Imposing the boundary conditions yields

$$A = C = \frac{\lambda_2 p_2}{\int_0^\infty e^{-\lambda_1 x} g(x) dx} = \frac{\lambda_2 p_2}{g^*(\lambda_1)} \tag{7.94}$$

and so $p_1(x)$ and $p_0(x)$ become

$$p_1(x) = \frac{\lambda_2 p_2 e^{-\lambda_1 x} [1 - G(x)]}{g^*(\lambda_1)}$$

and

$$p_0(x) = \frac{\lambda_2 p_2 \left(1 - e^{-\lambda_1 x} \right) [1 - G(x)]}{g^*(\lambda_1)} . \tag{7.95}$$

To find p_0, p_1, and p_2 we need to use the condition

$$p_2 + \int_0^\infty p_1(x) dx + \int_0^\infty p_0(x) dx = p_2 + p_1 + p_0 = 1 . \tag{7.96}$$

After some simplification, it can be verified that

$$p_2 = \frac{g^*(\lambda_1)}{g^*(\lambda_1) + \lambda_2 E(Y)} ,$$

$$p_1 = \frac{\lambda_2 [1 - g^*(\lambda_1)]}{\lambda_1 [g^*(\lambda_1) + \lambda_2 E(Y)]} , \tag{7.97}$$

$$p_0 = \frac{\lambda_2 [\lambda_1 E(Y) - 1 + g^*(\lambda_1)]}{\lambda_1 [g^*(\lambda_1) + \lambda_2 E(Y)]} ,$$

where $E(Y)$ is the mean length of time to repair a failed unit. The long-run system unavailability is given by $U = p_0$.

7.4 Problems and comments

Problems for Section 7.1

1. Consider any non-negative random variable T with c.d.f. $F(t)$. Suppose that $E(T) = \int_0^\infty t \, dF(t) < \infty$.
   (a) Show that $E(T) < \infty$ implies that $\lim_{t \to \infty} tR(t) = 0$, where $R(t) = 1 - F(t)$.
   (b) Show that $E(T) = \int_0^\infty R(t) dt$.
   (c) If T possesses finite second moment $(T^2) < \infty$ show that $E(T^2) = \int_0^\infty 2tR(t) dt$. More generally, if T possesses finite k'th moment $E(T^k) < \infty$, show that $E(T^k) = \int_0^\infty kt^{k-1} R(t) dt$.

2. Consider a 2-unit repairable system, both units active, with exponential failure and exponential repair distributions (model 2(a)). Suppose that $\lambda = .1/$hour and $\mu = 2/$hour. Plot $R(t)$ for $0 \le t \le 500$.

3. Plot $R(t)$ on the same axis system as before for the two-unit system in which one unit is active and the other is an inactive standby (model 2(b)). Compare the two models in terms of their reliability.

4. (a) Find the first passage time distribution into E_0 for the two-unit systems treated in Section 7.1, assuming that the system is initially in E_1 (one unit in a good state, the other unit in a failed state and under repair).

   (b) Find the mean first passage time from E_1 into E_0.

5. Suppose that we have two active units A and B, each initially in a good state and a single repairman. Times to failure are independent exponential with rates $\lambda_A = 1$ and $\lambda_B = 2$ for units A and B, respectively. The times to repair units A and B are exponential with rates $\mu_A = 6$ and $\mu_B = 10$, respectively. Repair of a unit is assumed to start as soon as the unit fails. System failure occurs when both units are down simultaneously. Find the first passage time distribution from E_2 (both units good) into E_0 (both units down). Find the mean first passage time from E_2 into E_0.

   Hint: There are four states in this system, the obvious E_0, E_2, then E_A (A good, B under repair) and E_B (B good, A under repair).

6. In Problem 6.1.2, starting in state E_2 let T be the first entrance time to state E_0. Find its p.d.f. and expectation.

7. Do the same to Problem 6.1.3.

Problems for Section 7.2

1. (a) Show that $\mathrm{Var}(T_{n0})$, the variance of the first passage time from $E_n \to E_0$ is given by $d^2/ds^2 (\log A(s))$ evaluated at $s = 0$, where $A(s)$ is given by Equation (7.55).

   (b) Show that $\mathrm{Var}(T_{n0}) = \sum_{i=1}^{n} s_i^{-2}$.

2. Given a three-unit system, all three units active, each having failure rate λ and one repair facility, with repair rate μ, compute $\ell_{30}, \ell_{31}, \ell_{32}, \ell_{20}, \ell_{21}$ and ℓ_{10}.

3. Suppose that there are three docks in a port, that the arrival rate for ships is .5 ships per day and that the servicing rate per dock is 1 ship per day. Let $E_i =$ state in which i ships are either being serviced or waiting for service. Thus E_4 is the state in which three ships are being serviced and one ship is waiting for service. Suppose that the initial state is E_0 (no ships being serviced). Find the mean number of days before we observe a ship waiting for service.

4. (a) For Example 1, find the exact and approximate values of $R_{30}(t)$ for $t = 10$ hours, $t = 100$ hours.

(b) Find approximate values for the time t_p such that there is probability p that the system will fail on or before time t_p. Do this for $p = .05, .10, .50, .90, .95$.

5. Consider a four-unit system with two repair facilities. Each unit is assumed to be active and has constant failure rate $\lambda = .1$/hour and each repair facility can repair a unit with constant repair rate $\mu = 1$/hour. The system is considered to be failed if 1 or fewer units are in a good state and operable if 2 or more units are in a good state. Using Equations (7.68) and (7.70), find the mean time to system failure (where time is measured from the instant that the system becomes operable until the instant it enters a failed state) and the mean time to repair the system (where time is measured from the instant that the system has entered a failed state until the instant that it enters an operable state).

6. Consider an n-unit repairable system, each unit active and having constant failure rate λ, and suppose that there are n repairmen, each capable of repairing a unit with repair rate μ. Suppose that a system is said to be in a good state if r units or more are operating properly and in a bad state if $(r-1)$ or fewer units are operating properly, $1 \leq r \leq n$. Substituting in Equations (7.69) and (7.71), show that the mean time to failure,

$$\ell_{r,r-1} = \frac{1+x}{\mu} \cdot \frac{1 - B\left(r - 1; n, \frac{x}{1+x}\right)}{nb\left(r - 1; n - 1, \frac{x}{1+x}\right)}$$

and that the mean time to repair,

$$\ell_{r-1,r} = \frac{1+x}{\mu} \cdot \frac{B\left(r - 1; n, \frac{x}{1+x}\right)}{nb\left(r - 1; n - 1, \frac{x}{1+x}\right)},$$

where $x = \mu/\lambda$; $b(k; n, p) = \binom{n}{k}p^k(1 - p)^{n-k}$ and $B(k; n, p) = \sum_{j=0}^{k} b(j; n, p)$.
Hint: For this problem $\lambda_j = j\lambda$, $j = 1, 2, \ldots, n$; $\mu_j = (n - j)\mu$, $j = 0, 1, 2, \ldots, n - 1$. See Problem 6.3.4.

7. Suppose that the total number of spares either on the shelf or on order for a certain part is two. The demand for this part is Poisson distributed with rate $\lambda = .1$/day. When a spare is taken from the shelf and placed in service, an order is placed immediately for a new spare. The time to fill this order is exponentially distributed with mean $\theta = 2$ days.

(a) Assuming that there are two spares on the shelf at $t = 0$, how many days will it take on the average until a demand for a spare cannot be filled?

(b) What is the long-run unavailability of a spare when needed?

Problems for Section 7.3

1. Using equations (7.86) and (7.87) find the Laplace transform of the p.d.f. of T, the first time to system failure, and $E(T)$ for a two-unit system—both units active and each unit having exponential time to failure with rate λ. There is a single repairman with time to repair distribution:

   (a) $g(t) = \mu^2 t e^{-\mu t}$, $t \geq 0$.

   (b) $g(t) = \alpha \mu_1 e^{-\mu_1 t} + (1 - \alpha)\mu_2 e^{-\mu_2 t}$, $t \geq 0$, $0 \leq \alpha \leq 1$.

   (c) $g(t) = \theta^{-1}$, $0 \leq t \leq \theta$.

2. Carry out the two examples and Problem 1 (above) for the two-unit case, where one unit is active, the second inactive until called upon for service.

3. Consider a two-unit system, both units active, where the time to failure of each unit is exponential with rate λ. There is a single repair man. Find the system unavailability if $\lambda = .01$ and time to repair is

   (a) Uniform in $[0, 2]$.

   (b) Constant 1.

   (c) Gamma distributed with both parameters equal to 2.

4. Consider the two-unit system in Section 7.3. Suppose that the system is in E_2 (both units good) at $t = 0$. Prove that

   (a) The time spent in E_2 before entering E_0 is exponential with rate $\lambda_2[1 - g^*(\lambda_1)]$.

   (b) The time spent in E_1 before entering E_0 is exponential with rate λ_1.

   (c) Using (a) and (b), verify that ℓ_{20}, the mean first passage time from E_2 into E_0, is given by Equation (7.87).

   Comment. If $p = 1 - g^*(\lambda_1)$ is very small, then it follows from (a) and (b) that the c.d.f. of T_{20}, the first passage time from E_2 into E_0, is closely approximated by $1 - e^{-\lambda_2 p t}$, $t \geq 0$ and $P(T_{20} > t) \approx e^{-\lambda_2 p t}$. For a proof of the result obtained in another way, see Gnedenko et al. [30].

5. (Continuation of Problem 4). Suppose that at $t = 0$, the system has just entered state $(1, x = 0)$. Show that

   (a) The time spent in E_1 before entering E_0 is exponential with rate λ_1.

   (b) The time spent in E_2 before entering E_0 is a random variable with c.d.f.

$$
\begin{array}{ll}
0, & t < 0 \\
p, & t = 0 \\
1 - q e^{-\lambda_2 p t}, & t > 0
\end{array}
$$

   where $q = 1 - p = g^*(\lambda_1)$.

   (c) Denote the mean time to pass from state $(1, x = 0)$ into the failed state by ℓ_{10}. Verify that $\ell_{10} = \ell_{20} - 1/\lambda_2$ (why is this obvious?).

6. Suppose that we define the random variable ξ_{01} as the length of time measured from the instant of entrance into the failed state until the instant of entrance into state $(1, x = 0)$. Let $\ell_{01} = E(\xi_{01})$. Show that

$$\ell_{01} = \frac{\lambda_1 E(Y) - 1 + g^*(\lambda_1)}{\lambda_1[1 - g^*(\lambda_1)]},$$

where $E(Y)$ is the mean length of time to repair a failed unit.

7. Define a cycle as the length of time measured from the instant of entrance into the failed state to the instant of next entrance into the failed state. Verify that the mean length of a cycle is $\ell_{01} + \ell_{10}$ and that the system unavailability $U = \ell_{01}/(\ell_{01} + \ell_{10})$.

8. Verify Equation (7.97) using the results of Problems 5, 6 and 7.

CHAPTER 8

Embedded Markov chains and systems reliability

8.1 Computations of steady-state probabilities

Suppose that we consider the continuous time stochastic process $N(t)$, the number of good units in an n-unit repairable system (with exponentially distributed unit times to failure and repair), only at those points in time when a change in state occurs, i.e., when a unit failure has just occurred or a unit repair has just been completed. We label the discrete set of time points $T_1 < T_2 < \cdots < T_m < \cdots$ at which the changes of state occur as $m = 1, 2, \ldots$, where m can be thought of as a discrete-time variable. We then define a discrete-time stochastic process with $n + 1$ possible states $0, 1, 2, \ldots, n$ by saying that the process is in state i at "time" m, if state E_i $(i = 0, 1, 2, \ldots, n)$ has just been entered into at T_m, the time of occurrence of the mth transition in the continuous-time process. The discrete-time process obtained in this way is said to be *embedded in the continuous-time process*. The initial state of this embedded process corresponds to the state of the system at $t = 0$, so that if we assumed, for example, that the system is initially in E_n (all units are good), then the state of the system at $m = 0$ is n. As a consequence of the Markovian character of the underlying continuous-time stochastic process (which arises from the assumption of exponentially distributed unit time to failure and repair distributions), it is not surprising that the sequence of states at $m = 0, 1, 2, \ldots$, for the embedded discrete-time Markov process forms what is called a *homogeneous Markov chain of the first order*. This means that the conditional probability of entering state j at the "time" point $m + 1$ given the states entered at "times" $0, 1, 2, \ldots, m$ is identically equal to the conditional probability of entering state j at "time" $m + 1$, given the state entered at "time" m. Thus, if the chain is in state i at "time" m, then the conditional probability of making the transition into state j at "time" $m + 1$ can be written simply as p_{ij}. The $(n + 1) \times (n + 1)$ matrix $\mathbf{P} = (p_{ij})$, $i, j = 0, 1, 2, \ldots, n$ is called the *transition probabilities matrix* of the Markov chain. If the initial state of the chain is specified at "time" $m = 0$, then any realization of the Markov chain at subsequent "times" $m = 1, 2, \ldots$ can be generated from knowledge of the matrix **P**. It is easy to compute the p_{ij}s if the underlying process in continuous-time is Markovian. For an introduction to Markov chains, see Feller [26]. Also, see, Cox and Miller [10] and Stroock [48]. Embedded Markov chains were introduced by D.G. Kendall [36],[37] in connection with the theory of queues.

It should be pointed out that Markov chains can also be embedded in continuous-time processes which are non-Markovian. Of particular importance in this connection are the so-called *semi-Markov processes*, which we shall soon have an opportunity to discuss. In this section, we shall show how one can compute steady-state probabilities of continuous-time Markov and semi-Markov processes by means of embedded Markov chains. In the second section, we shall see that mean first passage times for such processes can be obtained easily using embedded Markov chains.

We illustrate the concept of embedded Markov chains by several examples.

8.1.1 Example 1: One-unit repairable system

Consider a one-unit repairable system for which the c.d.f. of system failure (repair) times is $F_1(t)$ $(F_0(t))$ with mean $m_1(m_0)$. One way of describing the stochastic behavior of the system over time is by means of a random function $N(t)$, where $N(t) = 1(0)$ if the system is in the good (failed) state $E_1(E_0)$ at time t. A typical sample function of $N(t)$ involves the alternation of intervals when the system is in E_1 or E_0. The discrete-time Markov chain is associated with the discrete set of time-points at which transitions from E_1 to E_0 or E_0 to E_1 take place and the outcome $1(0)$ is assigned to the Markov chain at those time-points where a transition has just been made into $E_1(E_0)$. For any two successive time-points, the transition probabilities matrix for the Markov chain consisting of the states $0, 1$ is given by

$$\mathbf{P} = \begin{array}{c} \\ 0 \\ 1 \end{array} \begin{array}{c} \begin{array}{cc} 0 & 1 \end{array} \\ \left(\begin{array}{cc} 0 & 1 \\ 1 & 0 \end{array} \right) \end{array}.$$

It is customary to write to the left of the matrix a column of all the states, indicating transitions "from"; likewise, a row of all the states above the matrix, indicating transitions "into." If, at the mth transition point, state $1(0)$ of the embedded Markov chain has just been entered, then the $(m + 1)$'st transition will be into states $0(1)$ after a time interval drawn at random from the c.d.f. $F_1(t)$ $(F_0(t))$. If we assume, for convenience, that the system has just entered state E_1 at time $t = 0$, we can clearly generate sample random functions describing the stochastic behavior of the one-unit repairable system over time.

It should be noted that $N(t)$ is not a Markov process for all times t (where t is measured from the beginning of the process at $t = 0$) unless the c.d.f.s $F_i(t)$ are exponential. For general $F_i(t)$, $N(t)$ is not Markovian for all t, since one needs to know not only that the system is in state E_0 or E_1 at time t, but also how long it has been in this state. However, the process is Markovian at the discrete set of time points at which transitions are made from $E_0 \rightarrow E_1$ or $E_1 \rightarrow E_0$, since at these time points, we know not only the system state, but also that it has just entered this state.

Although the continuous-time process under discussion is not Markovian, it is a simple example of a class of continuous-time processes known as semi-Markov processes. A good way of describing a semi-Markov process is to show how a realization of a continuous-time process with $n+1$ states $E_0, E_1, E_2, \ldots,$ E_n can be constructed from a realization of a Markov chain with $n + 1$ states $0, 1, 2, \ldots, n$ and transition probability matrix $\mathbf{P} = (p_{ij})$, $i, j = 0, 1, 2, \ldots, n$. This is accomplished by the simple device of having the continuous-time process "spend" or "wait" an amount of time T_{ij}, where T_{ij} is a random variable with c.d.f. $F_{ij}(t)$, in state E_i (before making the transition into E_j) for each one-step transition from i to j in the Markov chain. As a result, $T_i = \min_{j \neq i} T_{ij}$ is the time spent in state E_i before making a transition. A continuous-time process generated in this manner is called an $(n + 1)$-state semi-Markov process corresponding to the pair of matrices $\mathbf{P}$ and $\mathbf{F} = (F_{ij}(t))$, $i, j = 0, 1, 2, \ldots, n$. If none of the $F_{ij}(t)$ are exponential, the semi-Markov process is Markovian only at those discrete points in time when transitions from one state to another take place (it is at these points in time that the Markov chain is embedded). If the $F_{ij}(t)$ are all exponential c.d.f.s, then the semi-Markov process becomes a Markov process.

8.1.2 Example 2: Two-unit repairable system

A two-unit repairable (Markov) system is given by its transitions and rates diagram as in Figure 8.1. Equivalently, the system is given by its transition

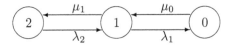

Figure 8.1 *Transitions and rates diagram of a two-unit repairable system.*

rates matrix Q,

$$Q = \begin{pmatrix} -\mu_0 & \mu_0 & 0 \\ \lambda_1 & -(\lambda_1 + \mu_1) & \mu_1 \\ 0 & \lambda_2 & -\lambda_2 \end{pmatrix}. \tag{8.1}$$

The matrix of transition probabilities for the discrete-time Markov chain embedded at the times when transitions in the continuous-time Markov process take place is given by

$$\mathbf{P} = \begin{matrix} & \begin{matrix} 0 & 1 & 2 \end{matrix} \\ \begin{matrix} 0 \\ 1 \\ 2 \end{matrix} & \begin{pmatrix} 0 & 1 & 0 \\ \frac{\lambda_1}{\lambda_1 + \mu_1} & 0 & \frac{\mu_1}{\lambda_1 + \mu_1} \\ 0 & 1 & 0 \end{pmatrix} \end{matrix}. \tag{8.2}$$

If the system has just entered state 0 (both units are bad) of the embedded Markov chain at the mth transition point, then it must go into state 1 (one unit is good, one unit is bad) at the $(m + 1)$'st transition point. Similarly, if the system has just entered state 2 (both units are good) at the mth transition point, then it must go into state 1 at the $(m + 1)$'st transition point. However, if the system has just entered state 1 at the mth transition point, either it will go into state 0 at the $(m+1)$'st transition point, if the one good unit fails before repair is completed on the one bad unit, or it will go into state 2 at the $(m + 1)$'st transition point if the one bad unit is repaired before the one good unit fails. It can be inferred from the transition probabilities that the time to failure of the good unit, when the system is in E_1 is an exponential random variable T_1 with rate λ_1, and that the time to repair the bad unit when the system is in E_1 is an exponential random variable Y with rate μ_1. Passage will occur from $1 \rightarrow 0$ if $T_1 < Y$ and from $1 \rightarrow 2$ if $T_1 > Y$. However,

$$\begin{aligned} P(T_1 < Y) &= \int_0^\infty \left\{ \int_{t_1}^\infty \lambda_1 \mu_1 e^{-(\lambda_1 t_1 + \mu_1 y)} dy \right\} dt_1 \\ &= \int_0^\infty \lambda_1 e^{-\lambda_1 t_1} e^{-\mu_1 t_1} dt_1 = \int_0^\infty \lambda_1 e^{-(\lambda_1 + \mu_1) t_1} dt_1 \\ &= \frac{\lambda_1}{\lambda_1 + \mu_1} = p_{10}. \end{aligned}$$

Similarly, $P(Y < T_1) = \mu_1/(\lambda_1 + \mu_1) = p_{12}$. Thus, the reader can see that the Matrix $\mathbf{P}$ is obtained from the matrix Q by first putting 0's in the diagonal entries of Q and then dividing each row by its sum.

The $F_{ij}(t)$ for this problem are:

$$
\begin{aligned}
F_{01}(t) &= 1 - e^{-\mu_0 t}, \quad t \geq 0 \\
F_{10}(t) &= F_{12}(t) = 1 - e^{-(\lambda_1 + \mu_1)t}, \quad t \geq 0
\end{aligned}
$$

and

$$
F_{21}(t) = 1 - e^{-\lambda_2 t}, \quad t \geq 0.
$$

Hence, the mean time spent in E_0 before making a transition from $E_0 \to E_1$ is $1/\mu_0$. The mean time spent in E_1 before making a transition to either E_0 or E_2 is $1/(\lambda_1 + \mu_1)$. The mean time spent in E_2 before making a transition from $E_2 \to E_1$ is $1/\lambda_2$.

8.1.3 Example 3: n-unit repairable system

An n-unit repairable system with transition probabilities from state to state in the time interval $(t, t + \Delta t]$ given in Table 6.3. The transition probabilities matrix $\mathbf{P}$ for the embedded Markov chain is given by

$$
\begin{array}{c}
\begin{array}{ccccccc}
& 0 & 1 & 2 & \cdots & \cdots & n-1 & n
\end{array} \\
\begin{array}{c}
0 \\ 1 \\ 2 \\ \vdots \\ n-1 \\ n
\end{array}
\left(
\begin{array}{ccccccc}
0 & 1 & 0 & \cdots & \cdots & 0 & 0 \\
\frac{\lambda_1}{\lambda_1+\mu_1} & 0 & \frac{\mu_1}{\lambda_1+\mu_1} & 0 & \cdots & 0 & 0 \\
0 & \frac{\lambda_2}{\lambda_2+\mu_2} & 0 & \frac{\mu_2}{\lambda_2+\mu_2} & 0 & 0 & 0 \\
0 & \cdots & \cdots & 0 & \frac{\lambda_{n-1}}{\lambda_{n-1}+\mu_{n-1}} & 0 & \frac{\mu_{n-1}}{\lambda_{n-1}+\mu_{n-1}} \\
0 & \cdots & \cdots & \cdots & 0 & 1 & 0
\end{array}
\right).
\end{array}
$$

More compactly, the elements of the transition probabilities matrix $\mathbf{P} \equiv (p_{ij})$ are given by:

$$
\begin{aligned}
p_{01} &= p_{n,n-1} = 1 \\
p_{i,i-1} &= \frac{\lambda_i}{\lambda_i + \mu_i}, \quad i = 1, 2, \ldots, n-1 \\
p_{i,i+1} &= \frac{\mu_i}{\lambda_i + \mu_i}, \quad i = 1, 2, \ldots, n-1
\end{aligned} \tag{8.3}
$$

and all other entries $= 0$.

As explained above, the matrix $\mathbf{P}$ is obtained from the transition rates matrix

$$
Q = \begin{pmatrix}
-\mu_0 & \mu_0 & 0 & \cdots & \cdots & 0 & 0 \\
\lambda_1 & -(\lambda_1 + \mu_1) & \mu_1 & 0 & \cdots & 0 & 0 \\
0 & \lambda_2 & -(\lambda_2 + \mu_2) & \mu_2 & 0 & 0 & 0 \\
& & & & & & \\
0 & \cdots & & \cdots & 0 & \lambda_{n-1} & -(\lambda_{n-1} + \mu_{n-1}) & \mu_{n-1} \\
0 & \cdots & & \cdots & \cdots & 0 & \lambda_n & -\lambda_n
\end{pmatrix}.
$$

The only proper c.d.f.s F_{ij} are given by:

$$
F_{01}(t) = 1 - e^{-\mu_0 t}, \qquad t \geq 0
$$

$$
F_{i,i-1}(t) = F_{i,i+1}(t) = 1 - e^{-(\lambda_i + \mu_i)t}, \quad t \geq 0, \quad i = 1, 2, \ldots, n-1
$$

and

$$
F_{n,n-1}(t) = 1 - e^{-\lambda_n t}, \qquad t \geq 0.
$$

The mean time spent in E_0 before a transition to E_1 is $1/\mu_0$. The mean time spent in E_i before a transition to either of its neighboring states E_{i-1} or E_{i+1}, $i = 1, 2, \ldots, n-1$, is $1/(\lambda_i + \mu_i)$, and the mean time spent in E_n before a transition to E_{n-1} is $1/\lambda_n$.

Remark 1 It is useful to introduce a few definitions at this point. For Markov chains with one step transition probabilities matrix $\mathbf{P} = (p_{ij})$, $p_{ij}^{(m)}$, the element in the ith row and j^{th} column of the matrix $\mathbf{P}^m$, is the probability that the chain is in state j after m transitions given that the chain was initially in state i. We say that state j is *accessible* from state i if state j can be reached from state i after a finite number of transitions, i.e., there exists an $m > 0$ such that $p_{ij}^{(m)} > 0$. A Markov chain is said to be *irreducible* if every state is accessible from every other state. If $p_{ii}^{(m)} = 0$, except for $m = d, 2d, 3d, \ldots$, state i is said to be *periodic* with period d. If $d = 1$, state i is said to be *aperiodic*. Another important concept is that of recurrence. Suppose that a Markov chain is initially in state i. The state i will be called *recurrent* if it is certain that the chain will ultimately return to i. If η_i, the mean number of steps to make a first return from $i \to i$, is finite (infinite), state i is said to be *positive-recurrent (null-recurrent)*. If the probability of ultimately returning to a starting state i is less than one, then state i will be called *transient*.

Of particular interest in this chapter are finite irreducible Markov chains (i.e., irreducible chains with a finite number of states). For such chains, it can be shown that all states are either aperiodic positive-recurrent, or periodic (with some common period d) positive-recurrent. An aperiodic positive-recurrent state is called *ergodic*.

Finite irreducible aperiodic Markov chains with transition probabilities matrix

$\mathbf{P} = (p_{ij})$, $i, j = 0, 1, 2, \ldots, n$ possess the property that $p_{ij}^{(m)}$, the probability that the chain is in state j after m steps, given that the initial state of the chain was i, tends to a limiting value, i.e., $\lim_{m \to \infty} p_{ij}^{(m)} = \pi_j$, independently of the initial state i. The steady-state probability distribution (also called *stationary distribution*) $\{\pi_j\}$ satisfies the system of linear equations

$$\pi_j = \sum_{i=0}^{n} \pi_i p_{ij}, \qquad j = 0, 1, 2, \ldots, n, \tag{8.4}$$

subject to the condition that $\sum_{j=0}^{n} \pi_j = 1$.

It is clear from Equation (8.4) why $\{\pi_0, \pi_1, \ldots, \pi_n\}$ is called a *stationary distribution*. What this means is that if the Markov chain starts out with $\boldsymbol{\pi} = (\pi_0, \pi_1, \ldots, \pi_n)$ as the initial probability distribution over the states $(0, 1, 2, \ldots, n)$, then the probability distribution over the states $(0, 1, 2, \ldots, n)$ after one step and indeed after any number of steps remains $\boldsymbol{\pi}$.

The empirical meaning of $\lim_{m \to \infty} p_{ij}^{(m)} = \pi_j$ is as follows: Consider a large number N of realizations of the finite irreducible aperiodic Markov chain, where each realization is initially in state i. Let $N_{ij}(m)$ be the number of these realizations which are in state j after m transitions have taken place. Then $N_{ij}(m)/N$, the empirical probability that the chain is in state j after m steps given that the initial state of the chain was i, will converge in probability to $p_{ij}^{(m)}$ as $N \to \infty$. As $m \to \infty$, $p_{ij}^{(m)}$ tends to the limiting value π_j. This is the justification for stating that π_j is the probability that the chain will be in state j, if we look at a realization of the process after a very large number of transitions have taken place.

The stationary distribution $\boldsymbol{\pi} = (\pi_0, \pi_1, \ldots, \pi_n)$ can be given another interpretation, which is very important in applications. Under this interpretation π_j, $j = 0, 1, 2, \ldots, n$ is the long-run proportion of times that the chain is in state j in an indefinitely long single realization of the chain. More precisely, let $X(M, j \mid i \text{ at } m = 0)$ be the number of times that the chain is in state j in the first M transitions of a realization of the chain, which is initially in state i. Then $X(M, j \mid i \text{ at } m = 0)/M$ will converge to π_j with probability one, independently of the initial state i (for a proof, see Cox and Miller [10]). This is the basis for the statement that if a large number M of transitions are observed, then approximately $M\pi_j$ of these will be into state j of the Markov chain. That $\boldsymbol{\pi} = (\pi_0, \pi_1, \pi_2, \ldots, \pi_n)$ is both an "ensemble" average taken over all realizations after a fixed (but very large) number of transitions and a "time" average taken over indefinitely long single realizations (at "times" $m = 0, 1, 2, \ldots$) is another example of what is meant by the term *ergodic*.

We are now in a position to write down an interesting and fundamental relationship between the steady-state probabilities $p_i = \lim_{t \to \infty} P_i(t)$, $i =$

$0, 1, 2, \ldots, n$ that an $(n + 1)$-state (continuous-time) Markov or semi-Markov process will be in state E_i and the stationary distribution $\boldsymbol{\pi} = (\pi_0, \pi_1, \ldots, \pi_n)$ of the Markov chain embedded in the continuous-time process. Suppose that the continuous-time process is observed until a large number M of system transitions take place. We have just seen that in this long realization, approximately $M\pi_i$ of the transitions will be into state i of the embedded Markov chain. Let us now assume that m_i is the mean time spent in state E_i of the continuous-time process in the course of a single transition from E_i to some other system-state. Clearly, then, the expected time spent in E_i associated with $M\pi_i$ visits to state i of the embedded Markov chain is given by $M\pi_i m_i$ and the total expected time associated with M transitions among states of the embedded Markov chain is $M\sum_{j=0}^{n} \pi_j m_j$. Hence, the long-run proportion of time spent by the continuous-time process in state E_i is approximately

$$\frac{M\pi_i m_i}{M\sum_{j=0}^{n} \pi_j m_j} = \frac{\pi_i m_i}{\sum_{j=0}^{n} \pi_j m_j}.$$

However, this is also the long-run probability $p_i = \lim_{t\to\infty} P_i(t)$ of finding the continuous-time process in state E_i at a point far out in time. Hence, we have

$$p_i = \frac{\pi_i m_i}{\sum_{j=0}^{n} \pi_j m_j}. \tag{8.5}$$

In particular, it can be shown for Example 3 that

$$\pi_1 = \left(1 + \frac{\mu_1}{\lambda_1}\right)\pi_0$$

$$\pi_2 = \left(\frac{\mu_1}{\lambda_1} + \frac{\mu_1\mu_2}{\lambda_1\lambda_2}\right)\pi_0$$

$$\pi_3 = \left(\frac{\mu_1\mu_2}{\lambda_1\lambda_2} + \frac{\mu_1\mu_2\mu_3}{\lambda_1\lambda_2\lambda_3}\right)\pi_0 \tag{8.6}$$

$$\vdots$$

$$\pi_{n-1} = \left(\frac{\mu_1\mu_2\cdots\mu_{n-2}}{\lambda_1\lambda_2\ldots\lambda_{n-2}} + \frac{\mu_1\mu_2\cdots\mu_{n-1}}{\lambda_1\lambda_2\ldots\lambda_{n-1}}\right)\pi_0$$

$$\pi_n = \frac{\mu_1\mu_2\cdots\mu_{n-1}}{\lambda_1\lambda_2\ldots\lambda_{n-1}}\pi_0.$$

where

$$\pi_0 = \frac{1}{2\left[1 + \frac{\mu_1}{\lambda_1} + \frac{\mu_1\mu_2}{\lambda_1\lambda_2} + \cdots + \frac{\mu_1\mu_2\ldots\mu_{n-1}}{\lambda_1\lambda_2\ldots\lambda_{n-1}}\right]}.$$

Furthermore,

$$\pi_0 m_0 = \pi_0 \frac{1}{\mu_0}$$

$$\pi_1 m_1 = \pi_0 \frac{1}{\lambda_1}$$

$$\pi_j m_j = \pi_0 \frac{\mu_1 \mu_2 \cdots \mu_{j-1}}{\lambda_1 \lambda_2 \cdots \lambda_j}, \qquad j = 2, \ldots, n.$$

Hence, using Equation (8.5), it follows that

$$p_0 = \frac{1}{1 + \frac{\mu_0}{\lambda_1} + \frac{\mu_0 \mu_1}{\lambda_1 \lambda_2} + \cdots + \frac{\mu_0 \mu_1 \cdots \mu_{n-1}}{\lambda_1 \lambda_2 \cdots \lambda_n}} \tag{8.7}$$

and

$$p_j = \frac{\mu_0 \mu_1 \mu_2 \cdots \mu_{j-1}}{\lambda_1 \lambda_2 \cdots \lambda_j} p_0,$$

which coincides with the result given in Equation (6.21).

Remark 2 In Remark 1 8.1.3, we limited our discussion to finite state, irreducible, aperiodic Markov chains. For an $(n + 1)$-state, irreducible, periodic Markov chain, there still exists a vector $\pi = (\pi_0, \pi_1, \ldots, \pi_n)$ which is the solution set of the system of Equations (8.4), subject to the condition that $\sum_{j=0}^n \pi_j = 1$. Thus, one can justifiably say that π is the stationary distribution for the periodic Markov chain, since if π is the initial probability distribution over the states $0, 1, 2, \ldots, n$, then this remains true following the first, second, and indeed any transition. However, unlike the aperiodic case, $\lim_{m \to \infty} p_{ij}^{(m)}$ does *not** converge to π_j (e.g., for a periodic chain with period d and $i = j$, $p_{jj}^{(m)} = 0$ except for $m = d, 2d, 3d, \ldots$). The fact that remains true for periodic as well as aperiodic chains is that if a large number M of transitions are observed, then approximately $M\pi_j$ of these will be into state j of the Markov chain. From this, it follows that if an $(n + 1)$-state Markov or semi-Markov process has embedded within it a periodic Markov chain, then the relationship (8.5) remains valid.

To illustrate what we have just said about periodic Markov chains, consider the two-state embedded Markov chain for the one-unit repairable system described in Example 1. The transition probabilities matrix for the embedded chain is periodic with period 2, since the chain must surely be in state 1(0) after an even number of transitions, given that it was initially in 1(0). One can compute formally the π_0 and π_1 associated with the transition probabilities matrix $\begin{pmatrix} 0 & 1 \\ 1 & 0 \end{pmatrix}$

---

* It is therefore wrong to say for a periodic chain that π_j is the long-run probability of finding the chain in state j. What does happen in the periodic case is that the sequence $p_{ij}^{(m)}$ is Cesaro-summable to π_j (independently of the initial state i) instead of actually converging to π_j. A sequence $a_1, a_2, \ldots$ is said to be Cesaro-summable to a limit b if the sequence $b_n = \frac{1}{n} \sum_{i=1}^n a_i$ converges to b.

using Equation (8.4), getting $\pi_0 = \pi_1 = 1/2$. The sequence $p_{11}^{(m)}$, which is $0, 1, 0, 1, 0, 1, \ldots$ certainly does not converge to $\pi_1 = 1/2$. The sequence $p_{11}^{(m)}$ is, however, Cesaro-summable to 1/2 since the sequence of averages $b_m = m^{-1} \sum_{i=1}^{m} p_{11}^{(m)} \rightarrow 1/2$ (note that if $m = 2n-1$, $b_m = (n-1)/(2n-1)$ and if $m = 2n$, $b_m = 1/2$). After $2M$ transitions have taken place, states 1(0) will have each been observed exactly $2M\pi_1(\pi_0) = M$ times. Of course, $\pi = (1/2, 1/2)$ is the stationary distribution for the chain with transition proba-bilities matrix $\begin{pmatrix} 0 & 1 \\ 1 & 0 \end{pmatrix}$. If $(1/2, 1/2)$ is the initial probability distribution over the states 0,1, then this remains true following any number of transitions.

Remark 3 The system of $(n+1)$ linear Equations (8.4) can be written in matrix form as $\pi \mathbf{P} = \pi$, where π is the row vector $(\pi_0, \pi_1, \ldots, \pi_n)$. Equivalently, Equations (8.4) can be written as $\pi (\mathbf{I}_{n+1} - \mathbf{P}) = \mathbf{0}_{n+1}$ where $\mathbf{I}_{n+1}$ is the $(n+1) \times (n+1)$ identity matrix and $\mathbf{0}_{n+1}$ is the $(n+1)$-component zero vector $(0, 0, \ldots, 0)$. It can be shown that

$$\pi_i = \frac{A_i}{\sum_{j=0}^{n} A_j}, \qquad i = 0, 1, 2, \ldots, n, \tag{8.8}$$

where A_i is the cofactor of the ii element of $\mathbf{I}_{n+1} - \mathbf{P}$, i.e., the cofactor obtained by striking out the ith row and the ith column of the matrix $\mathbf{I}_{n+1} - \mathbf{P}$.

Suppose that $\mathbf{P}$ is the transition probabilities matrix among states in a Markov chain embedded in a semi-Markov process and let m_i be the mean length of stay associated with each entrance into state i, $i = 0, 1, 2, \ldots, n$. It then follows using Equation (8.8) and substituting in Equation (8.5) that the long-run probability of the system being in state E_i for the continuous-time semi-Markov process is given by

$$p_i = \frac{A_i m_i}{\sum_{j=0}^{n} A_j m_j}, \qquad i = 0, 1, 2, \ldots, n. \tag{8.9}$$

As an illustration, the matrix $\mathbf{I} - \mathbf{P}$ for Example 2 is

$$\begin{pmatrix} 1 & -1 & 0 \\ \frac{-\lambda_1}{\lambda_1 + \mu_1} & 1 & \frac{-\mu_1}{\lambda_1 + \mu_1} \\ 0 & -1 & 1 \end{pmatrix}. \tag{8.10}$$

It is readily verified that

$$
A_0 = \begin{vmatrix} 1 & \frac{-\mu_1}{\lambda_1+\mu_1} \\ -1 & 1 \end{vmatrix} = \frac{\lambda_1}{\lambda_1 + \mu_1}
$$

$$
A_1 = \begin{vmatrix} 1 & 0 \\ 0 & 1 \end{vmatrix} = 1 \tag{8.11}
$$

$$
A_2 = \begin{vmatrix} 1 & -1 \\ \frac{-\lambda_1}{\lambda_1+\mu_1} & 1 \end{vmatrix} = \frac{\mu_1}{\lambda_1 + \mu_1}.
$$

Hence,

$$
\pi_0 = \frac{A_0}{A_0 + A_1 + A_2} = \frac{\lambda_1}{2(\lambda_1 + \mu_1)}
$$

$$
\pi_1 = \frac{A_1}{A_0 + A_1 + A_2} = \frac{1}{2} \tag{8.12}
$$

$$
\pi_2 = \frac{A_2}{A_0 + A_1 + A_2} = \frac{\mu_1}{2(\lambda_1 + \mu_1)}.
$$

The computation of p_0, p_1 and p_2 is left as a problem for the reader.

8.1.4 Example 4: One out of n repairable systems

[This example is due to George Weiss.] Consider an n-unit repairable system with exponential time to failure of rate λ_i for the ith unit. Suppose that the system fails if any one of the n units fails. Assume further that only one unit can fail at a time and that during the repair of a failed unit the $(n-1)$ unfailed units cannot fail. Assume that m_i is the mean time to repair the ith unit. Define the $n+1$ system-states $E_0, E_1, E_2, \ldots, E_n$ as follows:

E_0 — system is in a good state.
E_i — system is in a failed state due to a failure of the ith unit, $i = 1, 2, \ldots, n$.

It is readily verified that the transition probabilities matrix for the Markov chain embedded at the time points where the state of the system undergoes transitions is given by the matrix $\mathbf{P} =$

$$
\begin{array}{c}
 \\
0 \\
1 \\
2 \\
\vdots \\
i \\
\vdots \\
n
\end{array}
\begin{array}{cccccccc}
0 & 1 & 2 & \cdots & i & \cdots & n \\
\left(\begin{array}{ccccccc}
0 & \lambda_1/\Lambda & \lambda_2/\Lambda & \cdots & \lambda_i/\Lambda & \cdots & \lambda_n/\Lambda \\
1 & & & & & & \\
1 & & & & & & \\
\vdots & & & & & & \\
1 & & & 0 & & & \\
\vdots & & & & & & \\
1 & & & & & &
\end{array} \right)
\end{array}
$$

$$(8.13)$$

where $\Lambda = \sum_{i=1}^{n} \lambda_i$.

The stationary distribution (for the embedded process) $\boldsymbol{\pi}= (\pi_0, \pi_1, \ldots, \pi_n)$ satisfying the set of Equations (8.4) and $\sum_{i=0}^{n} \pi_i = 1$ is:

$$\pi_0 = \frac{1}{2}$$

and

$$\pi_i = \lambda_i/2\Lambda, \qquad i = 1, 2, \ldots, n. \tag{8.14}$$

The total exit rate from state E_0 is Λ, thus, the mean time spent in E_0 during a single visit is $1/\Lambda$ and the mean time spent in E_i during a single visit is m_i. Consequently, the steady-state probability that the system is in E_0 at a point far out in time is given by:

$$p_0 = \lim_{t \to \infty} P_0(t) = \frac{1}{2\Lambda} \left/ \left(\frac{1}{2\Lambda} + \frac{\sum_{j=1}^{n} \lambda_j m_j}{2\Lambda} \right) \right. = \frac{1}{1 + \sum_{j=1}^{n} \lambda_j m_j}.$$

$$(8.15)$$

Similarly, the steady-state probability that the system is in the failed state E_i at a point far out in time is given by:

$$p_i = \lim_{t \to \infty} P_i(t) = \frac{\lambda_i m_i}{1 + \sum_{j=1}^{n} \lambda_j m_j} \qquad i = 1, 2, \ldots, n. \tag{8.16}$$

In this example, p_0 is also the system availability, i.e., the long-run proportion of time spent by the system in the good state E_0.

8.1.5 Example 5: Periodic maintenance

There are many examples of systems whose availability, mean times to system failure, and similar characteristics can be increased enormously by maintenance and repair. This is particularly so if redundancy is built into the system so that failures at lower levels (e.g., subsystem, component, etc.) need not result in a system failure. For such systems, periodic maintenance can be used to increase the mean time to system failure and system availability. With this introduction, we now consider the next example.

It is assumed that the system we are considering has the time to failure c.d.f. $F(t)$, if it is not maintained. In order to increase its operating life, periodic maintenance is scheduled at times $T, 2T, \ldots$ At these times, failed or otherwise suspect components, subsystems, etc., are replaced so that the system can be considered as being restored to its original good state (as it was at $t = 0$) after the maintenance is completed. This goes on until the system fails at some time T_1 between two consecutive scheduled maintenance times kT and $(k+1)T$, where $k = 0, 1, 2, \ldots$. If, for simplicity, the time involved in performing maintenance is zero, then T_1 would be the time to the first system failure. If the system is restored to its original good state at T_1 and the above process is repeated (i.e., first scheduled maintenance is at $T_1 + T$, second scheduled maintenance is at $T_1 + 2T, \ldots$, etc.), then the next system failure will occur at $T_2 > T_1$, the third system failure at $T_3 > T_2$, etc. Under the assumption that the system is brought back to its original state either at the scheduled maintenance time points or at the times $\{T_i\}$ when system failure occurs, the random variables $T_1, T_2 - T_1, \ldots, T_{i+1} - T_i, \ldots$ will have a common distribution $F_T(t)$ and a common mean $\mu_T = E(T_1)$. The subscript T is used to denote the c.d.f. of system-life and the mean time to system failure associated with scheduled maintenance at times $T, 2T, \ldots$. We now find $R_T(t) = 1 - F_T(t)$, the reliability of the maintained system. It is verified that

$$
\begin{aligned}
R_T(t) &= R(t), & 0 \leq t < T \\
&= R(T)R(t - T), & T \leq t < 2T \\
&= [R(T)]^2 R(t - 2T), & 2T \leq t < 3T \\
&\ \ \vdots \\
&= [R(T)]^j R(t - jT), & jT \leq t < (j+1)T .
\end{aligned}
\tag{8.17}
$$

The mean time to system failure is given by

$$
\mu_T = \int_0^\infty R_T(t)dt = \sum_{j=0}^\infty \int_{jT}^{(j+1)T} R_T(t)dt .
\tag{8.18}
$$

But

$$
\begin{aligned}
\int_{jT}^{(j+1)T} R_T(t)dt &= [R(T)]^j \int_{jT}^{(j+1)T} R(t - jT)dt \\
&= [R(T)]^j \int_0^T R(\tau)d\tau .
\end{aligned}
\tag{8.19}
$$

Substituting into Equation (8.18), we obtain

$$
\begin{aligned}
\mu_T &= \sum_{j=0}^{\infty} [R(T)]^j \int_0^T R(\tau) d\tau \\
&= \frac{\int_0^T R(\tau) d\tau}{1 - R(T)} = \frac{\int_0^T [1 - F(\tau)] d\tau}{F(T)}.
\end{aligned}
\tag{8.20}
$$

We note that as $T \to \infty$, $R_T(\cdot) \to R(\cdot)$ and $\mu_T \to \mu$, where μ is the mean time to system failure when no periodic maintenance takes place. Another way of establishing Equation (8.20) is as follows:

Let the discrete random variable N denote the number of scheduled maintenance actions until first system failure occurs. If $N = j$, this means that first system failure occurs between jT and $(j+1)T$. Clearly,

$$
P(N = j) = [1 - F(T)]^j F(T), \qquad j = 0, 1, 2, \dots. \tag{8.21}
$$

If $N = j$, the observed amount of time until first system failure $= jT + \xi$, where ξ is a random variable with truncated c.d.f. $F(\tau)/F(T)$, $0 \le \tau \le T$. The expected amount of time until system failure occurs is, therefore, given by

$$
\begin{aligned}
\mu_T &= E(N)T + \frac{\int_0^T \tau dF(\tau)}{F(T)} \\
&= \left[\frac{1}{F(T)} - 1\right] T + \frac{TF(T) - \int_0^T F(\tau) d\tau}{F(T)} = \frac{\int_0^T [1 - F(\tau)] d\tau}{F(T)}.
\end{aligned}
$$

Up to this point, we have assumed that it takes no time to perform scheduled maintenance or unscheduled repair. Let us now assume that $\tau_m = $ average length of time to perform a scheduled maintenance and that $\tau_r = $ average length of time to perform an unscheduled repair. The length of time between the completion of a scheduled maintenance (or unscheduled repair) and the start of the next scheduled maintenance is assumed to be T.

Suppose that a cycle consists of the total time which elapses from the instant that the system leaves the failed state until the next exit from the failed state. The total time Z consists of three components:

(1) The time Z_1 spent by the system in an operable state.
(2) The time Z_2 spent in performing periodic maintenance in order to bring the system back to its original state.
(3) The time Z_3 spent in bringing the system back to its original state from a failed state.

Clearly,

$$E(Z) = E(Z_1) + E(Z_2) + E(Z_3).\qquad(8.22)$$

From Equations (8.20) and (8.15),

$$E(Z_1) = \int_0^T [1 - F(t)]dt/F(T) = \mu_T,$$

$$E(Z_2) = \left[\frac{1}{F(T)} - 1\right]\tau_m,$$

and finally

$$E(Z_3) = \tau_r.$$

Hence,

$$E(Z) = \frac{\int_0^T [1 - F(t)]dt}{F(T)} + \left[\frac{1}{F(T)} - 1\right]\tau_m + \tau_r.\qquad(8.23)$$

The system availability is defined as

$$A(T) = \frac{E(Z_1)}{E(Z)} = \frac{\int_0^T [1 - F(t)]dt}{\int_0^T [1 - F(t)]dt + [1 - F(T)]\tau_m + F(T)\tau_r}.\qquad(8.24)$$

If $A(T)$ is increasing, then it attains its maximum at $T^* = \infty$, and no scheduled maintenance is recommended. Otherwise, if $A(T)$ attains its maximum at a finite point T^*, this value is recommended for the periodic maintenance (see Problems 9 and 11). A recent book which deals with the statistical and management aspects of preventive maintenance is Gertsbakh [29].

Example 5(a). Consider a system consisting of n identical active units, where the time to failure p.d.f. for each unit is given by the exponential distribution $\lambda e^{-\lambda t}$, $t \geq 0$. Assume that the system fails if, and only if, all n units have failed. Scheduled maintenance involving the replacement of failed units by good units takes place at time points $T, 2T, 3T, \ldots$ until the system fails at some time between scheduled maintenance times. It is assumed that it takes no time to perform scheduled maintenance.

The unattended system reliability is $R(t) = 1 - (1 - e^{-\lambda t})^n$, $t \geq 0$. For this particular problem we can compute μ_T explicitly. The numerator of Equation (8.20) is equal to

$$\int_0^T [1 - (1 - e^{-\lambda t})^n]dt = T - S_n,$$

where

$$S_j = \int_0^T (1 - e^{-\lambda t})^j dt = \int_0^T (1 - e^{-\lambda t})^{j-1}dt - \int_0^T (1 - e^{-\lambda t})^{j-1}e^{-\lambda t}dt$$

$$= S_{j-1} - \frac{1}{\lambda j}(1 - e^{-\lambda T})^j, \qquad j = 2, 3, \dots, n$$

and

$$S_1 = T - \frac{1}{\lambda}(1 - e^{-\lambda T}).$$

It follows that

$$T - S_n = \frac{1}{\lambda} \sum_{j=1}^{n} \frac{1}{j}(1 - e^{-\lambda T})^j.$$

Thus, Equation (8.20) becomes here

$$\mu_T = \frac{1}{\lambda} \sum_{j=1}^{n} \frac{1}{j}(1 - e^{-\lambda T})^{j-n}.$$

If $n = 2$ and $\lambda T = .1$, $\mu_T = 11.008/\lambda$.

If $n = 3$ and $\lambda T = .1$, $\mu_T = 116.013/\lambda$.

These values should be compared with the mean time to failure of the nonmaintained system, which corresponds to choosing $T = \infty$, namely $\mu_\infty = 1.5/\lambda$ for $n = 2$ and $\mu_\infty = 1.833/\lambda$ for $n = 3$.

Steady-state distributions. It is instructive to compute system availability for the case of a system undergoing periodic maintenance of the kind described in Example 5, using the method of embedded Markov chains. The chain is embedded at those points in time when the system has just started scheduled maintenance, when it has just failed (prior to the next scheduled maintenance), or when it has completed either scheduled maintenance or unscheduled repair. We define the following states:

E_2: System is in its original good state or operable and not undergoing maintenance.
E_1: System is operable but is undergoing scheduled maintenance.
E_0: System is in a failed state and being repaired.

The transition probabilities matrix for the embedded Markov chain is

$$\mathbf{P} = \begin{array}{c} \\ 2 \\ 1 \\ 0 \end{array} \begin{array}{c} 2 \\ \left(\begin{array}{ccc} 0 & 1-F(T) & F(T) \\ 1 & 0 & 0 \\ 1 & 0 & 0 \end{array} \right) \end{array}$$

If the system has just entered E_2 at the nth transition point, either it will go into E_1 at the $n + 1^{\text{st}}$ transition point if it has not failed during the interval of length T before the next scheduled maintenance, or it will go into E_0 at the $(n + 1)^{\text{st}}$ transition point if it fails before the next scheduled maintenance. If the system has just entered E_1 or E_0 at the nth transition point, then it must enter E_2 at the $(n+1)^{\text{st}}$ transition point.

The stationary distribution $\boldsymbol{\pi} = (\pi_0, \pi_1, \pi_2)$ over the states of the embedded Markov chain is:

$$\pi_0 = \frac{F(T)}{2}, \quad \pi_1 = \frac{1 - F(T)}{2}, \quad \pi_2 = \frac{1}{2}. \tag{8.25}$$

The mean time spent in E_0 from the time of entrance into E_0 until the time of exit from E_0 is assumed to be $m_0 = \tau_r$. The mean time spent in E_1 from the time of entrance into E_1 until the time of exit from E_1 is assumed to be $m_1 = \tau_m$. The mean time spent in E_2 from the time of entrance into E_2 until the time of exit from E_2 is $m_2 = \int_0^T [1 - F(t)]dt$. Hence, the long-run probability that the system is in state E_2 at a point far out in time is:

$$p_2 = \frac{\pi_2 m_2}{\pi_0 m_0 + \pi_1 m_1 + \pi_2 m_2} = \frac{\int_0^T [1 - F(t)]dt}{\int_0^T [1 - F(t)]dt + [1 - F(T)]\tau_m + F(T)\tau_r}.$$

This agrees with Equation (8.24), which we obtained by other arguments. Because of ergodicity, p_2 is the long-run system availability.

8.1.6 Example 6: Section 7.3 revisited

In Section 7.3, we investigated the stochastic behavior over time of a two-unit repairable system, with exponentially distributed unit time to failure distributions and a general unit time to repair distribution, $G(y)$. The system is said to be in state E_i, $i = 0, 1, 2$, at time t, if there are i good units at time t. The steady-state probabilities p_0, p_1, p_2 of being in state E_i are given in Equation (7.97). In this example, we use embedded Markov chains to derive p_0, p_1, p_2. The Markov chain is embedded at those time-points when a state change occurs. If the system makes an entrance into state E_i, the corresponding state for the chain will be denoted by i. The transition probabilities matrix for the embedded Markov chain is given by

$$\mathbf{P} = \begin{array}{c} \\ 2 \\ 1 \\ 0 \end{array} \begin{array}{ccc} 2 & 1 & 0 \\ \left(\begin{array}{ccc} 0 & 1 & 0 \\ \lambda_1 G^*(\lambda_1) & 0 & 1 - \lambda_1 G^*(\lambda_1) \\ 0 & 1 & 0 \end{array} \right) \end{array}.$$

The entries in the first and third rows of the matrix are obvious since we must necessarily go from $2 \to 1$ and $0 \to 1$ in a single transition. To obtain the entries in the second row of the matrix, we note that from state 1, the transition is made either to state 2 or state 0 depending on whether the failed unit is repaired before or after the unfailed unit fails. The time to failure of the unfailed unit is denoted by a random variable T_1, which is distributed with p.d.f. $f_1(t) =$

$\lambda_1 e^{-\lambda_1 t}$, $t \geq 0$. The time to repair the failed unit is denoted by a random variable Y with c.d.f. $G(y)$ (and p.d.f. $g(y)$ if $G'(y)$ exists). The probability p_{10} of making the transition from $1 \rightarrow 0$ is given by

$$p_{10} = P(Y > T_1) = \int_0^\infty \lambda_1 e^{-\lambda_1 t}[1 - G(t)]dt$$

$$= 1 - \lambda_1 \int_0^\infty e^{-\lambda_1 t} G(t)dt. \qquad (8.26)$$

If we denote the Laplace transform of $G(y)$ as

$$G^*(s) = \int_0^\infty e^{-sy} G(y)dy, \qquad (8.27)$$

then Equation (8.26) becomes

$$p_{10} = P(Y > T_1) = 1 - \lambda_1 G^*(\lambda_1). \qquad (8.28)$$

Also,

$$p_{12} = P(Y < T_1) = 1 - p_{10} = \lambda_1 G^*(\lambda_1). \qquad (8.29)$$

This verifies the transition probabilities given in the second row of the matrix. The stationary distribution $\pi = (\pi_0, \pi_1, \pi_2)$ over the states in the embedded Markov chain is:

$$\pi_0 = \frac{1}{2}[1 - \lambda_1 G^*(\lambda_1)]$$

$$\pi_1 = \frac{1}{2} \qquad (8.30)$$

$$\pi_2 = \frac{1}{2}\lambda_1 G^*(\lambda_1).$$

We now compute the mean length of stay at each visit of the system to states $0, 1, 2$. The length of stay in each visit to state 2 of the embedded Markov chain is an exponential random variable X_2 with rate λ_2. Hence,

$$m_2 = E(X_2) = 1/\lambda_2. \qquad (8.31)$$

Similarly, the length of stay in each visit to state 1 of the embedded Markov chain measured from the time of entrance into E_1 until the time of exit from E_1 is a random variable X_1, which is the minimum of the failure time of the unfailed unit and the repair time of the failed unit. Thus, X_1 has c.d.f.

$$H(t) = P(X_1 \leq t) = 1 - e^{-\lambda_1 t}[1 - G(t)], \quad t \geq 0. \qquad (8.32)$$

Hence, m_1, the mean length of time associated with each visit to state 1, is

$$m_1 = E(X_1) = \int_0^\infty [1 - H(t)]dt = \int_0^\infty e^{-\lambda_1 t}[1 - G(t)]dt$$

$$= \frac{1}{\lambda_1} - G^*(\lambda_1). \qquad (8.33)$$

The reason why m_2 and m_1 were so easy to compute is that, at the instant of entrance into E_2 or into E_1, all past history becomes irrelevant. To compute m_0, the mean length of stay in the state when both units are down, is more difficult. The reason is that, at the instant of entrance into E_0, we must take into account the amount of repair time y that has been spent on the unit that is being repaired. It can be shown that the probability that the amount of repair time Y on the unit under repair lies between y and $y + dy$ at the moment of entrance into the failed state is given by

$$\eta(y)dy = \frac{\lambda_1 e^{-\lambda_1 y}[1 - G(y)]dy}{1 - \lambda_1 G^*(\lambda_1)}. \tag{8.34}$$

The conditional probability that the system is in the failed state for an additional length of time lying between z and $z + dz$, given that the amount of repair time prior to entrance into the failed state is y, is given by

$$dG(z + y)/[1 - G(y)]. \tag{8.35}$$

Consequently, the conditional expected additional time spent in the failed state is given by

$$
\begin{aligned}
E(Z \mid Y = y) &= \frac{\int_0^\infty z\,dG(z + y)}{1 - G(y)} = \frac{\int_y^\infty (w - y)dG(w)}{1 - G(y)} \\
&= \frac{E(Y) - y + \int_0^y G(w)dw}{1 - G(y)}.
\end{aligned} \tag{8.36}
$$

But, m_0, the unconditional mean time spent in the failed state, is given by

$$m_0 = \int_0^\infty E(Z \mid Y = y)\eta(y)dy. \tag{8.37}$$

Substituting Equations (8.34) and (8.36) into Equation (8.37) and simplifying yields

$$m_0 = \frac{E(Y) - \frac{1}{\lambda_1} + G^*(\lambda_1)}{1 - \lambda_1 G^*(\lambda_1)} = \frac{E(Y)}{1 - \lambda_1 G^*(\lambda_1)} - \frac{1}{\lambda_1}. \tag{8.38}$$

Knowing π_0, π_1, π_2 and m_0, m_1, m_2 we can now calculate the long-run probabilities p_0, p_1, p_2 that the system is in states E_0, E_1, E_2 after a long time has elapsed. In particular, p_0, the system unavailability U, becomes

$$U = p_0 = \frac{\pi_0 m_0}{\pi_0 m_0 + \pi_1 m_1 + \pi_2 m_2}. \tag{8.39}$$

Substituting Equations (8.30), (8.31), (8.33) and (8.38) into Equation (8.39) yields

$$U = p_0 = \frac{\lambda_2 \left[E(Y) - \frac{1}{\lambda_1} + G^*(\lambda_1)\right]}{\lambda_2 E(Y) + \lambda_1 G^*(\lambda_1)}. \tag{8.40}$$

Noting that $g^*(\lambda) = \lambda G^*(\lambda)$, this result is obtained by a different method in Equation (7.97).

8.1.7 Example 7: One-unit repairable system with prescribed on-off cycle

Consider a system which is required to operate according to a prescribed duty cycle, e.g., on for a length of time T_d and off for a length of time T_0, and that this pattern of operation is continued indefinitely. Suppose that during the duty interval, the system failure and repair rates are λ_d and μ_d, respectively. During the off-period, it is assumed that failures are impossible (so that $\lambda_0 = 0$) and that the repair rate is μ_0. We are interested in finding:

(1) The steady-state probability that the system is in a good state at the start of a duty interval. This is also called *operational readiness*.

(2) The steady-state probability that the system will be in a good state through-out a duty interval. This is also called *interval reliability*.

(3) The expected fraction of time that the system is in a good state during a duty interval. This is also called *interval availability*.

To solve this problem, we embed a Markov chain at the start of each duty interval. It is a two-state chain with state $1(0)$ corresponding to "system is in a good (bad) state" at the start of a duty interval. The transition probabilities matrix of the chain can be written as

$$\mathbf{P} = \begin{array}{c} \\ 1 \\ 0 \end{array} \begin{pmatrix} \overset{1}{A} & \overset{0}{1-A} \\ B & 1-B \end{pmatrix},$$

where A is the probability that the system is in a good state at the start of the $(m+1)^{\text{st}}$ duty interval, given that it was in a good state at the start of the mth duty interval. Similarly, B is the probability that the system is in a good state at the start of the $(m+1)^{\text{st}}$ duty interval, given that it was in a bad state at the start of the mth duty interval. Let $P_{ij}(T_d)$, $i = 0,1$; $j = 0,1$ represent the probability that the system is in state j at the end of a duty interval given that it was in state i at the start of the duty interval. Similarly, define $P_{ij}(T_0)$ for an interval when the system is turned off. Using the results in Chapter 4, we can write the following formulae for $P_{ij}(T_d)$:

$$P_{11}(T_d) = \frac{\mu_d}{\lambda_d + \mu_d} + \frac{\lambda_d}{\lambda_d + \mu_d} e^{-(\lambda_d+\mu_d)T_d}$$

$$P_{10}(T_d) = \frac{\lambda_d}{\lambda_d + \mu_d} - \frac{\lambda_d}{\lambda_d + \mu_d} e^{-(\lambda_d+\mu_d)T_d}$$

$$P_{01}(T_d) = \frac{\mu_d}{\lambda_d + \mu_d} - \frac{\mu_d}{\lambda_d + \mu_d} e^{-(\lambda_d+\mu_d)T_d} \qquad (8.41)$$

$$P_{00}(T_d) = \frac{\lambda_d}{\lambda_d + \mu_d} + \frac{\mu_d}{\lambda_d + \mu_d} e^{-(\lambda_d+\mu_d)T_d} .$$

For the off-interval, we have

$$P_{11}(T_0) = 1 \qquad\qquad P_{01}(T_0) = 1 - e^{-\mu_0 T_0}$$

$$P_{10}(T_0) = 0 \qquad\qquad P_{00}(T_0) = e^{-\mu_0 T_0} . \qquad (8.42)$$

It is readily verified that A and B can be expressed as follows in terms of the $P_{ij}(T_d)$ and $P_{ij}(T_0)$:

$$A = P_{11}(T_d) + P_{10}(T_d)P_{01}(T_0) \qquad (8.43)$$

and

$$B = P_{01}(T_d) + P_{00}(T_d)P_{01}(T_0) .$$

The steady-state distribution $\boldsymbol{\pi} = (\pi_0, \pi_1)$ for the embedded Markov chain is the solution set for the pair of equations

$$\pi_0 B + \pi_1 A = \pi_1$$

$$\pi_0 + \pi_1 = 1 . \qquad (8.44)$$

Hence, π_1, the steady-state probability that the embedded chain is in a good state at the start of a duty cycle (operational readiness), is

$$\pi_1 = B/(1 - A + B) . \qquad (8.45)$$

This is the answer to (1). The steady-state probability that the system will be in a good state throughout a duty interval (interval reliability) is given by $\pi_1 e^{-\lambda_d T_d}$. This is the answer to (2).

Finally, the expected fraction of time that the system is in a good state during a duty interval (interval availability) is given by

$$\text{Interval Availability} = \frac{1}{T_d} \int_0^{T_d} \{\pi_1 P_{11}(t) + \pi_0 P_{01}(t)\}\, dt . \qquad (8.46)$$

It can be verified that the right-hand side of (8.46) becomes

$$\frac{\mu_d}{\lambda_d + \mu_d} + [\pi_1(\lambda_d + \mu_d) - \mu_d] \left\{ \frac{1 - e^{-(\lambda_d+\mu_d)T_d}}{(\lambda_d + \mu_d)^2 T_d} \right\} .$$

8.2 Mean first passage times

Mean first passage times computed in Chapter 7 by differentiating the Laplace transform of the p.d.f. of the first passage time, can be obtained much more easily by using embedded Markov chains. We illustrate this by means of examples.

8.2.1 Example 1: A two-unit repairable system

A two-unit repairable system with exponentially distributed unit times to failure and repair. As before, E_i is the state in which i units are good, $i = 0, 1, 2$; E_0 is an absorbing state. The matrix of transition probabilities for the discrete Markov chain embedded at the times when transitions in the continuous-time process take place is given by

$$
\mathbf{P} = \begin{array}{c} 0 \\ 1 \\ 2 \end{array} \begin{pmatrix} \overset{0}{1} & \overset{1}{0} & \overset{2}{0} \\ \frac{\lambda_1}{\lambda_1 + \mu_1} & 0 & \frac{\mu_1}{\lambda_1 + \mu_1} \\ 0 & 1 & 0 \end{pmatrix}.
$$

Note that 0 is an absorbing state from which exit to any other state is impossible.

Let ℓ_{20} = mean time to first enter E_0 given that the system started in E_2 and let ℓ_{10} = mean time to first enter E_0 given that the system started in E_1. It is clear that the following equations are satisfied by ℓ_{20} and ℓ_{10}:

$$
\begin{aligned}
\ell_{20} &= \frac{1}{\lambda_2} + \ell_{10} \\
\ell_{10} &= \frac{1}{\lambda_1 + \mu_1} + \frac{\mu_1}{\lambda_1 + \mu_1} \ell_{20} .
\end{aligned} \tag{8.47}
$$

The first equation says that the mean first passage time into E_0 given that the system is initially in E_2 equals the mean time to leave E_2 in the first step (which takes the system into E_1) plus the mean first passage time from E_1 (our new starting point) into E_0. The second equation says that the mean first passage time into E_0 given that the system is initially in E_1 equals the mean time to leave E_1 in the first step (which takes the system either into the absorbing state E_0 with probability $\lambda_1/(\lambda_1 + \mu_1)$ or into state E_2 with probability $\mu_1/[\lambda_1 + \mu_1]$) plus the probability that the first step takes the system into state E_2 times the mean first passage time from E_2 (our new starting point) into E_0.

Solving for ℓ_{20} and ℓ_{10} yields:

$$\ell_{20} = \frac{1}{\lambda_2} + \frac{1}{\lambda_1}\left(1 + \frac{\mu_1}{\lambda_2}\right)$$

$$\ell_{10} = \frac{1}{\lambda_1}\left(1 + \frac{\mu_1}{\lambda_2}\right), \tag{8.48}$$

which coincides with results obtained in Chapter 7.

8.2.2 Example 2: General repair distribution

Same as Example 1, except that the repair distribution is described by a c.d.f. $G(y)$, with associated Laplace transform $G^*(s)$. Then the transition probabilities matrix of the Markov chain embedded at the times when transitions in the continuous-time process take place is given by

$$\mathbf{P} = \begin{array}{c} 0 \\ 1 \\ 2 \end{array} \begin{pmatrix} \overset{0}{1} & \overset{1}{0} & \overset{2}{0} \\ 1 - \lambda_1 G^*(\lambda_1) & 0 & \lambda_1 G^*(\lambda_1) \\ 0 & 1 & 0 \end{pmatrix}.$$

The pair of equations satisfied by ℓ_{20} and ℓ_{10} are:

$$\ell_{20} = \frac{1}{\lambda_2} + \ell_{10}$$

$$\ell_{10} = \left[\frac{1}{\lambda_1} - G^*(\lambda_1)\right] + \lambda_1 G^*(\lambda_1)\,\ell_{20}. \tag{8.49}$$

The first equation in (8.49) coincides with the first equation in (8.48). The first term on the right-hand side of the second equation in (8.49) is the mean time spent in E_1 from the instant of entrance into E_1 from E_2 until the instant of departure from E_1. This has been computed previously as m_1 in Equation (8.33). In the second term, the multiplier of ℓ_{20} is p_{12}, as given by Equation (8.29). Solving for ℓ_{20} gives

$$\ell_{20} = \frac{\frac{1}{\lambda_2} + \frac{1}{\lambda_1} - G^*(\lambda_1)}{1 - \lambda_1 G^*(\lambda_1)} = \frac{\lambda_1 + \lambda_2 - \lambda_1\lambda_2 G^*(\lambda_1)}{\lambda_1\lambda_2(1 - \lambda_1 G^*(\lambda_1))}. \tag{8.50}$$

This coincides with Equation (7.87) if we note that $g^*(\lambda_1) = \lambda_1 G^*(\lambda_1)$.

8.2.3 Example 3: Three-unit repairable system

In the three-unit repairable systems, with exponentially distributed unit times to failure and repair, E_i is the state in which i units are good, $i = 0, 1, 2, 3$ and

E_0 is an absorbing state. The transition probabilities matrix for the discrete Markov chain embedded at the times when transitions in the continuous-time process take place is given by

$$\mathbf{P} = \begin{array}{c} \\ 0 \\ 1 \\ 2 \\ 3 \end{array} \begin{pmatrix} \begin{array}{cccc} 0 & 1 & 2 & 3 \end{array} \\ \begin{array}{cccc} 1 & 0 & 0 & 0 \\ \frac{\lambda_1}{\lambda_1+\mu_1} & 0 & \frac{\mu_1}{\lambda_1+\mu_1} & 0 \\ 0 & \frac{\lambda_2}{\lambda_2+\mu_2} & 0 & \frac{\mu_2}{\lambda_2+\mu_2} \\ 0 & 0 & 1 & 0 \end{array} \end{pmatrix}.$$

It is readily verified that ℓ_{30}, ℓ_{20}, and ℓ_{10} satisfy the following set of linear equations:

$$\begin{aligned} \ell_{30} &= \frac{1}{\lambda_3} + \ell_{20} \\ \ell_{20} &= \frac{1}{\lambda_2 + \mu_2} + \frac{\mu_2}{\lambda_2 + \mu_2}\ell_{30} + \frac{\lambda_2}{\lambda_2 + \mu_2}\ell_{10} \qquad (8.51) \\ \ell_{10} &= \frac{1}{\lambda_1 + \mu_1} + \frac{\mu_1}{\lambda_1 + \mu_1}\ell_{20} \,. \end{aligned}$$

In vector-matrix notation, the set of Equations (8.51) can be written as follows:

$$\begin{pmatrix} \ell_{10} \\ \ell_{20} \\ \ell_{30} \end{pmatrix} = \begin{pmatrix} 0 & \frac{\mu_1}{\lambda_1+\mu_1} & 0 \\ \frac{\lambda_2}{\lambda_2+\mu_2} & 0 & \frac{\mu_2}{\lambda_2+\mu_2} \\ 0 & 1 & 0 \end{pmatrix} \begin{pmatrix} \ell_{10} \\ \ell_{20} \\ \ell_{30} \end{pmatrix} + \begin{pmatrix} \frac{1}{\lambda_1+\mu_1} \\ \frac{1}{\lambda_2+\mu_2} \\ \frac{1}{\lambda_3} \end{pmatrix}.$$

$$(8.52)$$

Let $\mathbf{L}$ = column vector with components ℓ_{10}, ℓ_{20}, ℓ_{30}.
Let $\mathbf{M}$ = column vector with components $1/(\lambda_1 + \mu_1)$, $1/(\lambda_2 + \mu_2)$, $1/\lambda_3$
(note that the components are, respectively, the mean times spent in E_1, E_2, E_3 before making a transition to another state).
Let $\mathbf{P}_{(0,0)}$ = matrix obtained by striking out the 0 row and 0 column of the matrix $\mathbf{P}$.

Then, Equation (8.52) can be written in vector-matrix notation as:

$$\mathbf{L} = \mathbf{P}_{(0,0)}\mathbf{L} + \mathbf{M}. \qquad (8.53)$$

Hence, the solution for $\mathbf{L}$ is

$$\mathbf{L} = [\mathbf{I}_3 - \mathbf{P}_{(0,0)}]^{-1}\mathbf{M}. \qquad (8.54)$$

Since

$$\mathbf{I}_3 - \mathbf{P}_{(0,0)} = \begin{pmatrix} 1 & -\frac{\mu_1}{\lambda_1+\mu_1} & 0 \\ -\frac{\lambda_2}{\lambda_2+\mu_2} & 1 & -\frac{\mu_2}{\lambda_2+\mu_2} \\ 0 & -1 & 1 \end{pmatrix} \qquad (8.55)$$

and its determinant $|\mathbf{I}_3 - \mathbf{P}_{(0,0)}|$ is equal to $\lambda_1\lambda_2/[(\lambda_1 + \mu_1)(\lambda_2 + \mu_2)]$, the inverse matrix can be computed to be

$$[I_3 - P_{(0,0)}]^{-1} = \begin{pmatrix} \frac{\lambda_1+\mu_1}{\lambda_1} & \frac{\mu_1(\lambda_2+\mu_2)}{\lambda_1\lambda_2} & \frac{\mu_1\mu_2}{\lambda_1\lambda_2} \\ \frac{\lambda_1+\mu_1)}{\lambda_1} & \frac{(\lambda_1+\mu_1)(\lambda_2+\mu_2)}{\lambda_1\lambda_2} & \frac{\mu_2(\lambda_1+\mu_1)}{\lambda_1\lambda_2} \\ \frac{\lambda_1+\mu_1)}{\lambda_1} & \frac{(\lambda_1+\mu_1)(\lambda_2+\mu_2)}{\lambda_1\lambda_2} & \frac{\lambda_1\lambda_2+\lambda_1\mu_2+\mu_1\mu_2}{\lambda_1\lambda_2} \end{pmatrix}.$$

(8.56)

Hence,

$$\mathbf{L} = \begin{pmatrix} \frac{\lambda_1+\mu_1}{\lambda_1} & \frac{\mu_1(\lambda_2+\mu_2)}{\lambda_1\lambda_2} & \frac{\mu_1\mu_2}{\lambda_1\lambda_2} \\ \frac{\lambda_1+\mu_1}{\lambda_1} & \frac{(\lambda_1+\mu_1)(\lambda_2+\mu_2)}{\lambda_1\lambda_2} & \frac{\mu_2(\lambda_1+\mu_1)}{\lambda_1\lambda_2} \\ \frac{\lambda_1+\mu_1}{\lambda_1} & \frac{(\lambda_1+\mu_1)(\lambda_2+\mu_2)}{\lambda_1\lambda_2} & \frac{\lambda_1\lambda_2+\lambda_1\mu_2+\mu_1\mu_2}{\lambda_1\lambda_2} \end{pmatrix} \begin{pmatrix} \frac{1}{\lambda_1+\mu_1} \\ \frac{1}{\lambda_2+\mu_2} \\ \frac{1}{\lambda_3} \end{pmatrix}$$

(8.57)

Expanding the right-hand side of Equation (8.57) gives

$$\begin{aligned} \ell_{10} &= \frac{1}{\lambda_1} + \frac{\mu_1}{\lambda_1\lambda_2} + \frac{\mu_1\mu_2}{\lambda_1\lambda_2\lambda_3} \\ &= \frac{1}{\lambda_1}\left(1 + \frac{\mu_1}{\lambda_2} + \frac{\mu_1\mu_2}{\lambda_2\lambda_3}\right), \\ \ell_{20} &= \frac{1}{\lambda_1} + \frac{\lambda_1+\mu_1}{\lambda_1\lambda_2} + \frac{\mu_2(\lambda_1+\mu_1)}{\lambda_1\lambda_2\lambda_3} \\ &= \frac{1}{\lambda_2}\left(1 + \frac{\mu_2}{\lambda_3}\right) + \frac{1}{\lambda_1}\left(1 + \frac{\mu_1}{\lambda_2} + \frac{\mu_1\mu_2}{\lambda_2\lambda_3}\right), \\ \ell_{30} &= \frac{1}{\lambda_1} + \frac{\lambda_1+\mu_1}{\lambda_1\lambda_2} + \frac{\lambda_1\lambda_2+\lambda_1\mu_2+\mu_1\mu_2}{\lambda_1\lambda_2\lambda_3} \\ &= \frac{1}{\lambda_3} + \frac{1}{\lambda_2}\left(1 + \frac{\mu_2}{\lambda_3}\right) + \frac{1}{\lambda_1}\left(1 + \frac{\mu_1}{\lambda_2} + \frac{\mu_1\mu_2}{\lambda_2\lambda_3}\right), \end{aligned}$$

(8.58)

a result obtained previously in Chapter 7.

8.2.4 Computations based on s_{jk}

It is instructive to compute mean first passage times into the failed state E_0 in still another way. This involves computing s_{jk}, the expected number of times that state k of the embedded Markov chain is occupied before entrance into state 0 given that the initial state of the chain is j. If m_k is the mean time spent in the associated system state E_k during a single occupancy of state k, then it is readily verified that ℓ_{j0}, the mean first passage time from $E_j \rightarrow E_0$, is given by

$$\ell_{j0} = \sum_{k=1}^{n} s_{jk} m_k, \qquad j = 1, 2, \ldots, n. \qquad (8.59)$$

To illustrate the computation of the s_{jk}, let us return to Example 1 in this section. The following equations are satisfied by $s_{22}, s_{12}, s_{21}, s_{11}$:

$$s_{22} = 1 + s_{12} \qquad\qquad s_{21} = s_{11}$$

$$\text{and} \qquad\qquad\qquad\qquad (8.60)$$

$$s_{12} = \frac{\mu_1}{\lambda_1 + \mu_1} s_{22} \qquad s_{11} = 1 + \frac{\mu_1}{\lambda_1 + \mu_1} s_{21}.$$

Solving for $s_{22}, s_{12}, s_{21}, s_{11}$ yields:

$$s_{22} = \frac{\lambda_1 + \mu_1}{\lambda_1}; \quad s_{12} = \frac{\mu_1}{\lambda_1}; \quad s_{21} = \frac{\lambda_1 + \mu_1}{\lambda_1}; \quad s_{11} = \frac{\lambda_1 + \mu_1}{\lambda_1}.$$
$$(8.61)$$

Also $m_2 = 1/\lambda_2$ and $m_1 = 1/(\lambda_1 + \mu_1)$. Hence, substituting in Equation (8.59) we obtain

$$\ell_{20} = s_{22} m_2 + s_{21} m_1 = \frac{\lambda_1 + \mu_1}{\lambda_1 \lambda_2} + \frac{\lambda_1 + \mu_1}{\lambda_1} \frac{1}{\lambda_1 + \mu_1} = \frac{1}{\lambda_2} + \frac{1}{\lambda_1}\left(1 + \frac{\mu_1}{\lambda_2}\right).$$
$$(8.62)$$

Similarly,

$$\ell_{10} = s_{12} m_2 + s_{11} m_1 = \frac{\mu_1}{\lambda_1 \lambda_2} + \frac{1}{\lambda_1} = \frac{1}{\lambda_1}\left(1 + \frac{\mu_1}{\lambda_2}\right). \qquad (8.63)$$

Equations (8.62) and (8.63) coincide with Equation (8.48).

For Example 2 in this section, it can be verified in a similar way that

$$s_{22} = s_{21} = s_{11} = \frac{1}{1 - \lambda_1 G^*(\lambda_1)} \quad\text{and}\quad s_{12} = \frac{\lambda_1 G^*(\lambda_1)}{1 - \lambda_1 G^*(\lambda_1)}. \qquad (8.64)$$

We have seen earlier that $m_2 = 1/\lambda_2$ and $m_1 = 1/\lambda_1 - G^*(\lambda_1)$. Hence, substituting in Equation (8.59), we obtain

$$\ell_{20} = s_{22} m_2 + s_{21} m_1 = \frac{1}{1 - \lambda_1 G^*(\lambda_1)}\left[\frac{1}{\lambda_2} + \frac{1}{\lambda_1} - G^*(\lambda_1)\right]$$
$$(8.65)$$

$$= \frac{\lambda_1 + \lambda_2 - \lambda_1 \lambda_2 G^*(\lambda_1)}{\lambda_1 \lambda_2 (1 - \lambda_1 G^*(\lambda_1))}.$$

This coincides with Equation (8.50).

For Example 3 in this section, it can be verified that the following equations hold:

$$s_{33} = 1 + s_{23}$$

$$s_{23} = \frac{\mu_2}{\lambda_2 + \mu_2} s_{33} + \frac{\lambda_2}{\lambda_2 + \mu_2} s_{13}$$

$$s_{13} = \frac{\mu_1}{\lambda_1 + \mu_1} s_{23}.$$

$$s_{32} = s_{22}$$

$$s_{22} = 1 + \frac{\mu_2}{\lambda_2 + \mu_2} s_{32} + \frac{\lambda_2}{\lambda_2 + \mu_2} s_{12} \tag{8.66}$$

$$s_{12} = \frac{\mu_1}{\lambda_1 + \mu_1} s_{22}.$$

$$s_{31} = s_{21}$$

$$s_{21} = \frac{\mu_2}{\lambda_2 + \mu_2} s_{31} + \frac{\lambda_2}{\lambda_2 + \mu_2} s_{11}$$

$$s_{11} = 1 + \frac{\mu_1}{\lambda_1 + \mu_1} s_{21}.$$

Solving for s_{ij}, $i, j = 1, 2, 3$ gives

$$s_{11} = s_{21} = s_{31} = \frac{\lambda_1 + \mu_1}{\lambda_1} \; ;$$

$$s_{12} = \frac{\mu_1(\lambda_2 + \mu_2)}{\lambda_1 \lambda_2} \; ; \quad s_{22} = s_{32} = \frac{(\lambda_1 + \mu_1)(\lambda_2 + \mu_2)}{\lambda_1 \lambda_2} \; ; \tag{8.67}$$

$$s_{13} = \frac{\mu_1 \mu_2}{\lambda_1 \lambda_2} \; ; \quad s_{23} = \frac{\mu_2(\lambda_1 + \mu_1)}{\lambda_1 \lambda_2} \; ; \quad s_{33} = \frac{\lambda_1 \lambda_2 + \lambda_1 \mu_2 + \mu_1 \mu_2}{\lambda_1 \lambda_2} \; .$$

Substituting in Equation (8.59) and recalling that $m_3 = 1/\lambda_3$, $m_2 = 1/(\lambda_2 + \mu_2)$, and $m_1 = 1/(\lambda_1 + \mu_1)$ we obtain

$$\ell_{30} = s_{33} m_3 + s_{32} m_2 + s_{31} m_1$$

$$= \frac{\lambda_1 \lambda_2 + \lambda_1 \mu_2 + \mu_1 \mu_2}{\lambda_1 \lambda_2 \lambda_3} + \frac{\lambda_1 + \mu_1}{\lambda_1 \lambda_2} + \frac{1}{\lambda_1} \; . \tag{8.68}$$

It is readily seen that Equation (8.68) coincides with ℓ_{30} given in Equation (8.59).

Let $\mathbf{S} = (s_{jk})$ be the 3×3 matrix of the s_{jk}. Then, the set of Equations (8.66) can be written as

$$\mathbf{S} = \mathbf{P}_{(0,0)}\mathbf{S} + \mathbf{I}_3 \; , \tag{8.69}$$

where $\mathbf{P}_{(0,0)}$ is the matrix which appears in Equation (8.52). The solution for

S is

$$\mathbf{S} = \left[\mathbf{I}_3 - \mathbf{P}_{(0,0)}\right]^{-1}. \tag{8.70}$$

This matrix was evaluated in Equation (8.56). The s_{jk} obtained from the matrix inversion are, of course, precisely the same as those given in Equation (8.67). We also note that Equation (8.59) can be written in matrix notation as:

$$\mathbf{L} = \mathbf{SM} = \left[\mathbf{I}_3 - \mathbf{P}_{(0,0)}\right]^{-1}\mathbf{M}. \tag{8.71}$$

This coincides with Equation (8.54).

Historical remarks

The idea of a semi-Markov process was apparently conceived independently and essentially simultaneously in 1954 by Lévy [40] and Smith [47]. Also, in 1954, Takács [49] introduced this process and applied it to a problem involving counters. Closely related to semi-Markov processes are Markov renewal processes. A number of basic results about Markov renewal processes are given in Pyke [44],[45]. His papers also contain an extensive bibliography.

The theory of semi-Markov processes has many applications and one of the first people to recognize this was G.H. Weiss [52], who applied it to a reliability problem. Other references dealing with applications of semi-Markov processes are: Chapter 5 of Barlow and Proschan [3] on machine-repair and system reliability problems; Weiss [53],[54] on maintenance and surveillance plans; Takács [51] and Fabens [25] on queueing and inventory problems; Weiss and Zelen [55] on a model applied to clinical trials of patients with acute leukemia undergoing experimental therapy. Also of interest is a two-part paper by Jewell [32],[33] on Markov renewal programming.

8.3 Problems and comments

Problems for Section 8.1

1. Consider the one-unit repairable system of Example 1 with m_0 = mean time to repair the system and m_1 = mean time to system failure. Show that the long-run probabilities p_0, p_1 of the system being in E_0, E_1 are given by $p_0 = m_0/(m_0 + m_1)$ and $p_1 = m_1/(m_0 + m_1)$, respectively.
2. Verify Formulae (8.6) and (8.7) first for the case of a two-unit repairable system and then for an n-unit repairable system.
3. (a) Show that the embedded Markov chain for the two-unit repairable system discussed in Example 2 is periodic with period 2.
   (b) Show that η_i, the mean number of steps to make a first return from $i \to i$, is given by $\eta_0 = 2(\lambda_1 + \mu_1)/\lambda_1$, $\eta_1 = 2$, $\eta_2 = 2(\lambda_1 + \mu_1)/\mu_1$.

(c) Using the results of (b), find π_0, π_1, π_2 for the embedded chain.

(d) Given that the embedded chain is initially in state i, what is the probability that it will be in states $0, 1, 2$ after an odd (even) number of transitions? Solve for $i = 0, 1, 2$.

4.(a) Show that the embedded Markov chain for the n-unit repairable system discussed in Example 3 is periodic with period 2.

(b) For the case of $n = 3$, find the long-run probability of finding the chain in states $0, 1, 2, 3$ after a very large number of odd (even) transitions, given that the chain is initially in state i. Solve for $i = 0, 1, 2, 3$.

(c) Given that the embedded chain is initially in state i, find η_i, the mean number of steps to make a first return from $i \to i$.

(d) Show that the π_js in Equation (8.6) satisfy the condition $\Sigma_k \pi_{2k} = \Sigma_k \pi_{2k+1} = \frac{1}{2}$.

5. Compute the $\{\pi_0, \pi_1, \ldots, \pi_n\}$ in Equation (8.4) of the embedded Markov chain in Example 3 using the co-factors of the diagonal elements in the $(n+1) \times (n+1)$ matrix **I-P**.

6.(a) Show that the embedded chain in Example 4 is periodic with period 2.

(b) Find η_i, the mean number of steps to make a return from $i \to i$.

(c) Given that the embedded chain is initially in state i, what is the probability that the chain is in state $0, 1, 2, \ldots, n$ after a very large number of odd (even) transitions?

7. The steady-state probabilities in the numerical example of Section 5.2, for the states E_0, E_1, E_2, are 1/25, 6/25 and 18/25, respectively. Show that for the embedded Markov chain, the steady-state probabilities are 1/14, 1/2 and 3/7.

8. Consider a system consisting of a single active unit plus an inactive spare, which can be switched on automatically if the active unit fails. Assume that a unit has constant failure rate λ when active and failure rate zero, when inactive. Assume perfect switching. System failure is said to occur if a unit failure occurs and no spare is available as a replacement. Scheduled maintenance involving instantaneous replacement of a failed unit by a good unit takes place at time points $T, 2T, 3T, \ldots$ until the system fails at some time between maintenance points. Compute μ_T, the mean time to system failure when the above maintenance schedule is followed. What is the value of μ_T if $\lambda = .01$ and $T = 10$?

9. Do the same problem for a system consisting of a single active unit plus two inactive spares. All units are assumed to have constant failure rate λ, when active.

10. Show that if F is the exponential distribution in Equation (8.24), then $A(T)$ is increasing to a limit. Find this limit and argue whether or not a periodic maintenance is recommended.

11. Using Formula (8.24), compute A, if $F(t) = 1 - e^{-\lambda t} - \lambda t e^{-\lambda t}$, $t \geq 0$
    $\lambda = .001/$hour, $T = 100$ hours, $\tau_m = 1$ hour, $\tau_r = 5$ hours.

12. Show that if the hazard function $h(t) = f(t)/(1 - F(t))$ is increasing and
    $\tau_r > \tau_m$, then there exists a unique $T^* < \infty$ which maximizes system
    availability A (if $\tau_r \leq \tau_m$, then $T^* = \infty$, i.e., one should not schedule
    maintenance). Show that T^* is the solution of the equation

    $$h(T) \int_0^T [1 - F(t)]dt - F(T) = \frac{\tau_m}{\tau_r - \tau_m}$$

    and verify that the maximum value of $A = 1/[1 + h(T^*)(\tau_r - \tau_m)]$.

13. It has been assumed in Example 5 that maintenance is scheduled peri-
    odically at times $T, 2T, \ldots$. A more general policy involves using a ran-
    dom maintenance schedule, where the time to the next maintenance mea-
    sured from the time of completion of the preceding maintenance or repair
    (whichever occurred) is a random variable with c.d.f. $G(t)$. Assume that at
    $t = 0$, the system is new and let $F(t)$ be the time to failure c.d.f. of the
    unattended system.

    (a) Show that the mean time to first system failure (assuming that it takes no
        time to perform maintenance) is equal to

        $$\mu_G = \int_0^\infty [1 - F(t)][1 - G(t)]dt \Big/ \int_0^\infty F(t)dG(t).$$

    (b) Suppose that $\tau_m(\tau_r)$ is the mean time to return the system to its original
        state if maintenance (repair) is performed. Show that the formula for
        system availability analogous to Equation (8.24) is

        $$A = \frac{\mu_G \int_0^\infty F(t)dG(t)}{(\mu_G + \tau_r) \int_0^\infty F(t)dG(t) + \tau_m \int_0^\infty G(t)dF(t)}.$$

    (c) Show that Equations (8.20) and (8.23) can be obtained by assuming that

        $$G(t) = 0, \quad t < T$$
        $$= 1, \quad t \geq T.$$

14. An item, whose length of life T is a random variable with c.d.f. $F(t)$ and
    finite mean μ, is placed in service at time $t = 0$. At an increasing sequence
    of times $\{x_k\}$, $k = 1, 2, \ldots$, the item is checked to see whether or not it is
    still in a good state (x_0 is assumed to be equal to zero). If the item is found
    to be in a good state, it is continued in service. This is kept up until one of
    the checks reveals that the item has failed.

    (a) Let $Y =$ the time until the item fails and this failure is detected, assuming
        that the act of checking is perfect and that it takes no time to perform a

check. Show that

$$E(Y) = \sum_{k=0}^{\infty} x_{k+1}\{F(x_{k+1}) - F(x_k)\}\,.$$

(b) Let $Z =$ the amount of time the item failure remains undetected. Show that

$$E(Z) = \sum_{k=0}^{\infty} \int_{x_k}^{x_{k+1}} (x_{k+1} - t)dF(t)\,.$$

(c) Show that $E(Y) = \mu + E(Z)$.

15. (Continuation of Problem 14.) Suppose that the time between checks is a constant equal to x, so that the sequence of times at which checks take place is $\{x_k = kx\}$, $k = 1, 2, \dots$. Such a checking procedure is said to be a *period-checking procedure with period x*.

   (a) Show that if the item length of life follows an exponential distribution with rate λ, then the expected length of time until the item fails *and* failure is detected (at the next checking time) is $x/(1 - e^{-\lambda x})$.

   (b) Suppose that it takes no time at all to replace a failed item with a new item and that a periodic checking procedure with period x is applied also to the replacement items. Assuming that an item is available for use when in a good state and unavailable for use when in an undetected failed state, show that the long-run availability of the item and its replacements $= (1 - e^{-\lambda x})/x$.

   (c) What is the long-run availability if $\lambda = .001/\text{hour}$ and $x = 100$ hour?

16. Using the embedded chain approach, find the availability of a system, where the assumption of periodic maintenance is replaced by the more general assumption that intervals measured from the completion of the most recent maintenance or repair until the next scheduled maintenance are independent random variables with c.d.f. $G(t)$.

17. Verify that $E(Z \mid Y = y) = 1/\mu$ when Y is exponential with rate μ, in Example 6.

18. Prove Equation (8.34).

19. Carry out the details for Equation (8.36).

20. Find the long-run probabilities p_i that the system is in state E_i, $i = 1, 2$, using the formula $p_i = \pi_i m_i/(\pi_0 m_0 + \pi_1 m_1 + \pi_2 m_2)$.

21. Suppose that we describe the stochastic behavior of the system in Example 6 by the random function $N(t) = i$, if the system is in state E_i at time t, $i = 0, 1, 2$.

   (a) Is $N(t)$ a Markov process?

   (b) Is $N(t)$ a semi-Markov process?

22. Verify the statement following Equation (8.46).

23. Consider a system which is on for 10 hours and off for 1 hour and assume that this pattern is repeated indefinitely. The failure rate is $\lambda_d = .05/\text{hour}$ during the duty-interval and zero during the off-interval. The repair rate in both the duty and off intervals is $\mu_d = \mu_0 = 1/\text{hour}$. Find the operational readiness, interval reliability, and interval availability for the system.

Problems for Section 8.2

1. Solve the set of linear Equations (8.51) and verify that the solution coincides with the one given in Chapter 7.

2. Consider a Poisson process with constant event rate λ. Use the method of embedded Markov chains to show that the mean time until one observes two events which are closer together than τ time units is equal to

$$\frac{2 - e^{-\lambda\tau}}{\lambda(1 - e^{-\lambda\tau})}.$$

Compare to Problem 7.3.2, Example 2.

3. Suppose that we have a two-unit system—unit 1 having constant failure rate λ_1 and unit 2 having constant failure rate λ_2. Suppose that the repair times for units 1 and 2 are, respectively, exponential with rates μ_1, μ_2. Suppose that both units are initially in a good state. Show, using the method of embedded Markov chains, that the mean time until both units are in a failed state is given by

$$\frac{(\lambda_2 + \mu_1)(\lambda_1 + \mu_2) + \lambda_1(\lambda_1 + \mu_2) + \lambda_2(\lambda_2 + \mu_1)}{\lambda_1\lambda_2(\lambda_1 + \lambda_2 + \mu_1 + \mu_2)}.$$

If $\mu_1, \mu_2 \gg \lambda_1, \lambda_2$ show that this mean first passage-time into the failed state is approximately

$$\frac{1}{\lambda_1\lambda_2\left(\frac{1}{\mu_1} + \frac{1}{\mu_2}\right)}.$$

Hint: Let E_3 be the state when both units are good; E_2 the state when unit 2 is good and unit 1 is failed; E_1 the state when unit 1 is good and unit 2 is failed; E_0 the state when both units are failed.

4. Suppose that we have a two-unit system—unit 1 having constant failure rate λ_1 and unit 2 having constant failure rate λ_2. Suppose further that it takes a constant length of time τ_1 to repair unit 1 and a constant length of time τ_2 to repair unit 2. Suppose that both units are initially in a good state. Show, using the method of embedded Markov chains, that the mean time until both

units are in a failed state is given by:

$$\frac{1 + \frac{\lambda_1}{\lambda_2}\left(1 - e^{-\lambda_2 \tau_1}\right) + \frac{\lambda_2}{\lambda_1}\left(1 - e^{-\lambda_1 \tau_2}\right)}{\lambda_1\left(1 - e^{-\lambda_2 \tau_1}\right) + \lambda_2\left(1 - e^{-\lambda_1 \tau_2}\right)}.$$

Show also that if $\lambda_1 \tau_2$ and $\lambda_2 \tau_1$ are each $\ll 1$, then the mean time until both units are in a failed state is approximately

$$\frac{1}{\lambda_1 \lambda_2 (\tau_1 + \tau_2)}.$$

5. (Continuation of Problem 4.) Show that the mean time until both units are in a failed state, if the time to repair c.d.f.s of units 1 and 2 are, respectively, $G_1(y)$ and $G_2(y)$, $y \geq 0$, is given by:

$$\frac{1 + \lambda_1\left[\frac{1}{\lambda_2} - G_1^*(\lambda_2)\right] + \lambda_2\left[\frac{1}{\lambda_1} - G_2^*(\lambda_1)\right]}{\lambda_1\left[1 - G_1^*(\lambda_2)\right] + \lambda_2\left[1 - G_2^*(\lambda_1)\right]},$$

where $G_i^*(s) = \int_0^\infty e^{-st} G_i(t) dt$.

6. In Example 1, let the random variable ξ_{jk}, $j, k = 1, 2$, be the number of times that state E_k is occupied before entrance into state E_0 given that the initial state of the system is E_j.

   (a) Find the probability distribution of ξ_{jk} and verify that $E(\xi_{jk}) = s_{jk}$ as given in Equation (8.61).

   (b) Verify that $\text{Var}(\xi_{jk}) = (\lambda_1 + \mu_1)\mu_1/\lambda_1^2$, for $j, k = 1, 2$.

   (c) Let Y_{20} be the first passage time from E_2 into E_0. Find $\text{Var} Y_{20}$.

7. Consider a semi-Markov process with associated embedded Markov chain consisting of states $\{i\}$, $i = 0, 1, 2, \ldots, n$ with transition probabilities matrix $\mathbf{P} = (p_{ij})$ and with mean times m_i spent in each visit to state i.

   (a) Suppose that 0 is an absorbing state. Let ℓ_{i0} be the mean first passage time into E_0 given that the system is initially in E_i. Let $s_{ij}^{(0)}$ be the mean number of visits to E_j before entering E_0 given that the system is initially in E_i. Verify that

$$\ell_{i0} = m_i + \sum_{j=1}^n p_{ij}\ell_{j0}, \qquad i = 1, 2, \ldots, n$$

$$s_{ij}^{(0)} = \delta_{ij} + \sum_{k=1}^n p_{ik} s_{kj}^{(0)}, \qquad \text{where } \delta_{ij} = 1, \quad \text{if } i = j$$
$$= 0, \quad \text{if } i \neq j.$$

   Write these equations in matrix form.

8. Similar to Problem 7, except that the absorbing state is E_k. Give equations for ℓ_{ik} and $s_{ij}^{(k)}$ where $i \neq k$ and $j \neq k$.

9. Similar to Problem 7, except that the states E_k, where the subscript k belongs to some subset A of the integers $0, 1, 2, \ldots, n$ are absorbing states. Give equations for ℓ_{iA} and $s_{ij}^{(A)}$, where i and j are not in A.

10. Using the same notations as in Problem 7, suppose that all states in the embedded Markov chain communicate with each other. Let ℓ_{ij} be the mean time for first passage into state E_j given that the system is initially in E_i. Here ℓ_{ii} is considered as the mean recurrence time between successive entrances into E_i. Verify that

$$\ell_{ij} = m_i + \sum_{k \neq j} p_{ik} \ell_{kj}$$

and that

$$\ell_{ii} = \frac{m_i}{p_i} = \frac{\sum_{j=0}^{n} \pi_j m_j}{\pi_i}.$$

Here $\{\pi_j\}$ is the stationary distribution for the embedded Markov chain (see Equation (8.4)) and p_i is the long-run probability of finding the semi-Markov process in state E_i.

11. Consider the n-unit repairable system of Example 3 of Subsection 8.1.3 with its transition probabilities matrix. Verify that

$$\ell_{nn} = \frac{1}{\lambda_n} + \ell_{n-1,n}$$

$$\ell_{ii} = \frac{1}{\lambda_i + \mu_i} + \frac{\lambda_i}{\lambda_i + \mu_i} \ell_{i-1,i} + \frac{\mu_i}{\lambda_i + \mu_i} \ell_{i+1,i}, \quad i = 1, \ldots, n-1$$

$$\ell_{00} = \frac{1}{\mu_0} + \ell_{10}.$$

Show that $\ell_{ii} = m_i / p_i$, where $m_i = 1/(\lambda_i + \mu_i)$.
Hint: Use Formulae (7.69), (7.71) and (6.21).

CHAPTER 9

Integral equations in reliability theory

9.1 Introduction

In this chapter we make use of the fact that some system reliability problems give rise in a natural way to *regenerative stochastic processes.*

A stochastic process is said to be *regenerative* if it has embedded within it a sequence of time points called *points of regeneration.* A point τ is said to be a point of regeneration of a stochastic process $X(t)$ if the (conditional) distribution of $X(t)$ for $t > \tau$, given $X(t)$ for all $t \leq \tau$, is identical with the (conditional) distribution of $X(t)$ for $t > \tau$, given $X(\tau)$. It is this basic property which is used in setting up the integral equations for the examples in this chapter. For a discussion of regenerative processes, see Kalashnikov [34]. Points of regeneration are conceptually akin to what are called *recurrent events* (Chapter 13 of Feller [26]).

The points of regeneration in each of the seven examples are the times at which a change of state of the system occurs. We begin the chapter by discussing, in Example 1, a stochastic process generated by the successive replacement times of an item, which possesses general time to failure c.d.f. $F(t)$ and which is replaced instantaneously upon failure by a new item, etc., *indefinitely.* The stochastic process so generated is called a *renewal process.* This process is a prototype "par excellence" of a regenerative process, with the renewal times as points of regeneration. As a matter of fact, a characteristic feature of a regenerative process is that one can find one or more renewal processes embedded within it. This can be verified easily in each of the examples discussed in this chapter.

9.2 Example 1: Renewal process

9.2.1 Some basic facts

Consider items whose length of life is distributed with a general c.d.f. $F(t)$. At time $t = 0$, one of these items is placed in use (it is assumed that this item is new, i.e., has age zero). The item fails at T_1, say, and is replaced immediately by a second item (also new). At some later time $T_2 > T_1$ the second item fails and is replaced by a third item (also new), whose life is distributed with c.d.f. $F(t)$, etc. Thus, we generate times $T_1 < T_2 < \cdots < T_n < \cdots$ at which replacements (or renewals) take place, where the item life times $T_1, T_2 - T_1, T_3 - T_2, \ldots, T_n - T_{n-1}, \ldots$ are mutually independent random variables with common c.d.f. $F(t)$. The sequence of times $T_1, T_2, \ldots, T_n, \ldots$ is said to be a *renewal process* and we are particularly interested in the random variable $N(t)$, the number of renewals in $(0, t]$. It should be noted parenthetically that if the common c.d.f. of item lifetimes is exponential with rate λ, then $N(t)$ is a Poisson process with occurrence rate λ. For a discussion of renewal theory, see Cox [9], Chapter 3 of Barlow and Proschan [3], Chapters 6 and 11 of Feller [27], and Prabhu [43].

Let us denote $P[N(t) = k]$ as $P_k(t)$. Then it is readily verified that the following relations hold:

$$P_0(t) = 1 - F(t) \tag{9.1}$$

$$P_k(t) = \int_0^t P_{k-1}(t - \tau)dF(\tau), \quad k \geq 1, \tag{9.2}$$

$$= P[T_k \leq t] - P[T_{k+1} \leq t].$$

To obtain Equation (9.1), we use the fact that the event "no renewals in $(0, t]$" is identical with the event "the original item lives longer than time t." To obtain Equation (9.2), we observe that the event "exactly k renewals occur in $(0, t]$" will occur if:

(i) The original item fails at some time τ, $0 < \tau < t$.
(ii) Exactly $(k - 1)$ replacement items fail (i.e., $(k - 1)$ renewals occur) in the interval $(\tau, t]$, given that the original item failed at time τ.

Since F is the c.d.f. of τ, the failure time of the original item, and $P_{k-1}(t - \tau)$ is the probability of (ii), we obtain Equation (9.2) by using a continuous version of the theorem of total probability. Let us define the following Laplace transforms,

$$
\begin{aligned}
P_k^*(s) &= L\{P_k\}(s) = \int_0^\infty e^{-st} P_k(t)\, dt \\
F^*(s) &= L\{F\}(s) = \int_0^\infty e^{-st} F(t)\, dt \\
f^*(s) &= \int_0^\infty e^{-st} dF(t) = sF^*(s)\,.
\end{aligned}
\tag{9.3}
$$

Here $f^*(s)$ is called the Laplace-Stieltjes transform of $F(t)$ and is the Laplace transform of the p.d.f. $f(t)$ if $f(t) = F'(t)$ exists. It is furthermore easy to verify that $f^*(s) = sF^*(s)$. Taking Laplace transforms of both sides of Equations (9.1) and (9.2) and using the fact that the Laplace transform of the convolution of two functions is the product of the Laplace transforms of the functions, yields

$$
\begin{aligned}
P_0^*(s) &= \frac{1}{s} - F^*(s) = \frac{1 - f^*(s)}{s} \\
P_k^*(s) &= P_{k-1}^*(s) f^*(s)\,, \qquad k = 1, 2, \dots\,.
\end{aligned}
\tag{9.4}
$$

Using the second equation recursively yields

$$
P_k^*(s) = P_0^*(s)[f^*(s)]^k = \frac{[f^*(s)]^k - [f^*(s)]^{k+1}}{s}\,.
\tag{9.5}
$$

The time to the kth renewal T_k, is a sum of k independent random variables drawn from the common c.d.f. $F(t)$. The c.d.f. of T_k is denoted by $F_k(t)$ and the p.d.f. of T_k, when it exists, is denoted by $f_k(t)$. It is readily verified that the Laplace transform of $f_k(t)$ (or equivalently and more generally the Laplace-Stieltjes transform of $F_k(t)$) is given by $[f^*(s)]^k$ and hence the Laplace transform of $F_k(t)$ is given by $[f^*(s)]^k/s$. Taking inverse transforms of Equation (9.5) yields

$$
P_k(t) = F_k(t) - F_{k+1}(t)\,, \qquad k = 1, 2, \dots\,,
\tag{9.6}
$$

a result obtained by a direct probabilistic argument in Equation (1.46). For $k = 0$, we have already seen that $P_0(t) = 1 - F(t)$.

Of particular interest is

$$
M(t) = E[N(t)] = \sum_{k=0}^\infty k P_k(t) = \sum_{k=1}^\infty F_k(t)
\tag{9.7}
$$

the mean number of renewals in $(0, t]$. Taking the Laplace transform of both sides of Equation (9.7) yields

$$M^*(s) = L\{M\}(s) \quad = \quad \frac{1}{s}\sum_{k=1}^{\infty}[f^*(s)]^k$$

$$= \quad \frac{f^*(s)}{s[1 - f^*(s)]} = \frac{F^*(s)}{1 - f^*(s)} \cdot \qquad (9.8)$$

Simplifying Equation (9.8) yields

$$M^*(s) = F^*(s) + M^*(s)f^*(s) . \qquad (9.9)$$

Inverting Equation (9.9) gives the integral equation

$$M(t) = F(t) + \int_0^t M(t - \tau)dF(\tau) = \int_0^t [1 + M(t - \tau)]dF(\tau) . \quad (9.10)$$

This equation is called the *renewal equation for $M(t)$*. If the p.d.f. $f(t) = F'(t)$ exists, then differentiating both sides of Equation (9.10) with respect to t gives the integral equation

$$m(t) = f(t) + \int_0^t m(t - \tau)f(\tau)d\tau , \qquad (9.11)$$

where $m(t) = M'(t)$ is called the *renewal density* (or rate). It is readily verified, taking Laplace transforms of both sides of Equation (9.11) and defining $m^*(s) = \int_0^\infty e^{-st}m(t)dt$, that

$$m^*(s) = f^*(s) + m^*(s)f^*(s) . \qquad (9.12)$$

Hence,

$$m^*(s) = \frac{f^*(s)}{1 - f^*(s)} = \sum_{k=1}^{\infty}[f^*(s)]^k . \qquad (9.13)$$

Inverting Equation (9.13) gives $m(t) = \sum_{k=1}^{\infty} f_k(t)$, a result that could have been obtained directly from Equation (9.7). It is clear from this formula that $m(t)\Delta t$ (for small Δt) is the probability that a renewal occurs in $(t, t + \Delta t]$.

9.2.2 Some asymptotic results

The behavior of $N(t)$, $M(t)$ and $m(t)$ as $t \to \infty$ is of particular interest. If $F(t)$ has finite mean μ, it can be shown that $N(t)/t$ converges to $1/\mu$ with probability one. A direct consequence of this result is the so-called *elementary renewal theorem*, which states that $\lim_{t\to\infty} M(t)/t = 1/\mu$. If the time between renewals possesses a p.d.f. $f(t) = F'(t)$, satisfying certain modest regularity conditions, it can be shown that $\lim_{t\to\infty} m(t) = 1/\mu$.

If the time between renewals possesses not only finite mean μ, but also finite variance σ^2, it can be shown that $N(t)$ obeys a *central limit theorem* with mean t/μ and variance $\sigma^2 t/\mu^3$, i.e.,

$$\lim_{t \to \infty} P\left\{ \frac{N(t) - \frac{t}{\mu}}{\sqrt{\sigma^2 t/\mu^3}} \leq y \right\} = \int_{-\infty}^{y} \frac{1}{\sqrt{2\pi}} e^{-x^2/2} dx = \Phi(y). \qquad (9.14)$$

It can also be shown that as $t \to \infty$,

$$M(t) = \frac{t}{\mu} + \frac{\sigma^2 - \mu^2}{2\mu^2} + o(1).$$

Note that this says that for any distribution for which $\sigma = \mu$, $M(t) = t/\mu + o(1)$ as $t \to \infty$. A member of this class of distributions is the exponential distribution. For an exponential distribution with mean μ, $M(t) = t/\mu$ exactly for all $t \geq 0$.

One of the ways of obtaining the asymptotic behavior of $M(t)$ and $m(t)$ is to use some results from the theory of Laplace transforms, which state that the behavior of $M(t)$ and $m(t)$ as $t \to \infty$ can be determined from the behavior of $M^*(s)$ and $m^*(s)$ as $s \to 0$. In view of the fact that $L\{t^{\alpha-1}\}(s) = \int_0^\infty e^{-st} t^{\alpha-1} dt = \Gamma(\alpha)/s^\alpha$, $\alpha > 0$, one might hope that if $g^*(s) = L\{g(t)\}$ $(s) = \int_0^\infty e^{-st} g(t) dt$, then $\lim_{s \to 0} s^\alpha g^*(s) = C$ implies $\lim_{t \to \infty} g(t)/t^{\alpha-1} = C/\Gamma(\alpha)$. The exact regularity condition under which such a result can be found in Korevaar [38].

To illustrate the use of this method in determining the behavior of $m(t)$ and $M(t)$ as $t \to \infty$, let us assume that the time between renewals is distributed with p.d.f. $f(t)$ and possesses finite mean μ (and the regularity condition). Under this assumption, for small s,

$$f^*(s) = \int_0^\infty f(t)[1 - st + o(s)] dt = 1 - \mu s + o(s)$$

and hence, substituting in Equations (9.8) and (9.13) it is easily verified that $\lim_{s \to 0} s^2 M^*(s) = \lim_{s \to 0} sm^*(s) = 1/\mu$. Hence, provided that certain regularity conditions usually satisfied in applications are met, it follows that $\lim_{t \to \infty} M(t)/t = 1/\mu$ and $\lim_{t \to \infty} m(t) = 1/\mu$. If the additional assumption is made that the time between renewals has finite variance, σ^2, then the expansion of $f^*(s)$ in the neighborhood of $s = 0$, becomes

$$f^*(s) = 1 - \mu s + \frac{\sigma^2 + \mu^2}{2} s^2 + o(s^2).$$

Substituting in Equation (9.8) gives

$$M^*(s) = \frac{1}{\mu s^2} + \frac{\sigma^2 - \mu^2}{2\mu^2 s} + o\left(\frac{1}{s}\right).$$

Hence, we can say that

$$\lim_{s \to 0} s \left\{ M^*(s) - \frac{1}{\mu s^2} \right\} = \frac{\sigma^2 - \mu^2}{2\mu^2}.$$

This implies that

$$\lim_{t \to \infty} \left\{ M(t) - \frac{t}{\mu} \right\} = \frac{\sigma^2 - \mu^2}{2\mu^2}.$$

9.2.3 More basic facts

It is interesting to ask the following question about the renewal process. Suppose that the process is observed at time t. What is the probability that no failure occurs in the time interval $(t, t + \tau]$?

The event "no failure in the time interval $(t, t + \tau]$" can occur in the following mutually exclusive ways:

(i) The original item does not fail in the interval $(0, t + \tau]$.

(ii) Exactly $k (\geq 1)$ replacements of the original item are made in $(0, t]$ with the kth replacement occurring at time $u < t$, and surviving for a length of time exceeding $t + \tau - u$.

The probability of event (i) is $1 - F(t + \tau)$ and the probability of event (ii) is $\sum_{k=1}^{\infty} \int_0^t [1 - F(t + \tau - u)] dF_k(u)$. Denoting the probability of no failure in $(t, t + \tau]$ as $R(\tau; t)$ it follows that

$$
\begin{aligned}
R(\tau; t) &= 1 - F(t + \tau) + \sum_{k=1}^{\infty} \int_0^t [1 - F(t + \tau - u)] dF_k(u) \\
&= 1 - F(t + \tau) + \int_0^t [1 - F(t + \tau - u)] dM(u), \quad (9.15)
\end{aligned}
$$

since $M(u) = \sum_{k=1}^{\infty} F_k(u)$.

Equation (9.15) is only one of many probability statements that could be made regarding the item in use at time t. Let us define A_t, B_t, and C_t, respectively, as the age, remaining life, and total life length of this item. Clearly, $R(\tau; t) = P(B_t > \tau)$. In a manner analogous to that used in establishing Equation (9.15), one can show that the c.d.f.s of A_t, B_t, and C_t are given, respectively, by

$$
\begin{aligned}
P(A_t \leq x) &= \int_{t-x}^t \{1 - F(t - u)\} dM(u), \quad \text{if } x < t \\
&= 1, \quad\quad\quad\quad\quad\quad\quad\quad\quad\quad \text{if } x \geq t \quad (9.16)
\end{aligned}
$$

$$P(B_t \le y) = F(t + y) - F(t)$$

$$+ \int_0^t \{F(t + y - u) - F(t - u)\} dM(u) \qquad (9.17)$$

$$P(C_t \le z) = \int_0^z \{M(t) - M(t - x)\} dF(x), \qquad \text{if } z < t$$

$$= F(z)[1 + M(t)] - M(t), \qquad \text{if } z \ge t. \quad (9.18)$$

Assuming the existence of the p.d.f. $f(t)$ and renewal density $m(t)$, one can also describe the probability distributions of A_t, B_t, and C_t as follows:

The probability distribution of A_t has a continuous part and a discrete part, where the continuous part of A_t is distributed with density function

$$g_{A_t}(x) = m(t - x)[1 - F(x)] \quad \text{for } x < t \qquad (9.19)$$

and where a discrete probability mass $1 - F(t)$ is assigned to the point $A_t = t$, i.e., $P(A_t = t) = 1 - F(t)$.

The p.d.f. of B_t is given by

$$g_{B_t}(y) = f(t + y) + \int_0^t m(t - u) f(y + u) du, \qquad t \ge 0 \qquad (9.20)$$

and the p.d.f. of C_t is given by

$$g_{C_t}(z) = \begin{cases} f(z)\{M(t) - M(t - z)\}, & z < t \\ f(z)\{1 + M(t)\}, & z \ge t. \end{cases} \qquad (9.21)$$

Of particular interest are the limiting distributions of A_t, B_t, and C_t as $t \to \infty$. Assume for simplicity that we have a renewal process for which $\lim_{t \to \infty} m(t) = 1/\mu$ exists. Taking the limit as $t \to \infty$ in Equations (9.19), (9.20) and (9.21) leads formally to $\lim_{t \to \infty} g_{A_t}(x) = [1 - F(x)]/\mu$, $\lim_{t \to \infty} g_{B_t}(y) = [1 - F(y)]/\mu$, and $\lim_{t \to \infty} g_{C_t}(z) = z f(z)/\mu$, as the limiting p.d.f.s of A_t, B_t, and C_t. For a rigorous proof of these results, see, e.g. Prabhu [43].

9.3 Example 2: One-unit repairable system

Consider a one-unit repairable system with time to failure c.d.f. $F(t)$, $t \ge 0$ and time to repair c.d.f. $G(t)$. The system is said to be in state $E_1(E_0)$ if it is good (bad). Let $P_{ij}(t)$, $i, j = 0, 1$ be the probability that the system is in state E_j at time t, given that it has just entered E_i at time $t = 0$.

It is readily verified that the following equations hold for the $P_{ij}(t)$:

$$P_{11}(t) = [1 - F(t)] + \int_0^t P_{01}(t - x)dF(x)$$

$$P_{01}(t) = \int_0^t P_{11}(t - x)dG(x)$$

$$P_{10}(t) = \int_0^t P_{00}(t - x)dF(x) \tag{9.22}$$

$$P_{00}(t) = 1 - G(t) + \int_0^t P_{10}(t - x)dG(x).$$

The key fact used in deriving Equations (9.22) is that the times of failure and the times of repair completion are points of regeneration. To verify the first equation, we note that the event "system is in E_1 at time t given that it just entered E_1 at time $t = 0$" will occur if:

(a) The system is failure-free over $(0, t]$,

or

(b) The first system failure occurs at time x, $0 < x < t$ (thus passing from E_1 into E_0) and is in state E_1 at time t given that it just entered E_0 at time x.

The probability of the event (a) is $1 - F(t)$. As for event (b), X, the time of the first failure, is distributed with c.d.f. $F(x)$ and $P_{10}(t - x)$ is the probability that the system will be in E_1 at time t given that it just entered E_0 at time x. Integrating over all possible values of x gives $\int_0^t P_{01}(t - x)dF(x)$ as the probability of the event (b). Adding the probabilities of the mutually exclusive events (a) and (b) gives the first equation. Similarly, we can readily verify the remaining three equations in (9.22). Taking Laplace transforms of both sides of Equations (9.22) yields

$$P_{11}^*(s) = \frac{1 - f^*(s)}{s} + P_{01}^*(s)f^*(s)$$

$$P_{01}^*(s) = P_{11}^*(s)g^*(s)$$

$$P_{10}^*(s) = P_{00}^*(s)f^*(s) \tag{9.23}$$

$$P_{00}^*(s) = \frac{1 - g^*(s)}{s} + P_{10}^*(s)g^*(s),$$

where the transforms in (9.23) are defined as:

$$P_{ij}^*(s) = \int_0^\infty e^{-st}P_{ij}(t)dt$$

$$f^*(s) = \int_0^\infty e^{-st}dF(t) \tag{9.24}$$

$$g^*(s) = \int_0^\infty e^{-st}dG(t).$$

It is readily verified that the $P_{ij}^*(s)$ are

$$
\begin{aligned}
P_{11}^*(s) &= \frac{1 - f^*(s)}{s[1 - f^*(s)g^*(s)]} \\
P_{01}^*(s) &= \frac{g^*(s)[1 - f^*(s)]}{s[1 - f^*(s)g^*(s)]} \\
P_{10}^*(s) &= \frac{f^*(s)[1 - g^*(s)]}{s[1 - f^*(s)g^*(s)]} \\
P_{00}^*(s) &= \frac{1 - g^*(s)}{s[1 - f^*(s)g^*(s)]} \, .
\end{aligned}
\tag{9.25}
$$

It is possible, at least in principle, to compute $P_{ij}(t)$ by taking inverse transforms. For some $F(t)$ and $G(t)$ this inversion can be carried out with ease. Suppose, for example, that $F(t) = 1 - e^{-\lambda t}$, $t \geq 0$ and $G(t) = 1 - e^{-\mu t}$, $t \geq 0$. Then $f^*(s) = \lambda/(s + \lambda)$ and $g^*(s) = \mu/(s + \mu)$. Substituting in the first equation of (9.25) gives

$$
P_{11}^*(s) = \left(\frac{\mu}{\lambda + \mu} \right) \frac{1}{s} + \left(\frac{\lambda}{\lambda + \mu} \right) \frac{1}{s + \lambda + \mu} \, .
\tag{9.26}
$$

Inverting Equation (9.26) gives

$$
P_{11}(t) = \frac{\mu}{\lambda + \mu} + \frac{\lambda}{\lambda + \mu} e^{-(\lambda + \mu)t} \, ,
\tag{9.27}
$$

which coincides with Equation (4.8).

9.4 Example 3: Preventive replacements or maintenance

In this section we reconsider Example 5 of Subsection 8.1.5. The main interest is the effect of *preventive replacements* or *maintenance* on the time between failures of an item.

At time $t = 0$, an item whose time to failure is a nonnegative random variable distributed with c.d.f. $F(t)$ is placed in use. The item is replaced instantaneously by a new item either at the time of failure of the item, X_1, or at a planned replacement time Y_1, whichever comes first. The planned replacement time Y_1 is a nonnegative random variable distributed with c.d.f. $G(t)$. At $Z_1 = \min(X_1, Y_1)$ the process starts over again. The length of time between the time of removal of the first item and the time of removal of the second item is $Z_2 = \min(X_2, Y_2)$, where X_2 and Y_2 are independent random variables distributed with c.d.f.s $F(t)$ and $G(t)$, respectively. More generally, the length of time between the time of removal of the $(i - 1)^{st}$ item and the time of removal of the ith item is $Z_i = \min(X_i, Y_i)$, where X_i and Y_i are independent random variables distributed with c.d.f.s $F(t)$ and $G(t)$, respectively.

Clearly, the times at which removals take place form a renewal process, for which the time between successive renewals (removals) are independent observations drawn from the common c.d.f. $H(t) = 1 - [1 - F(t)][1 - G(t)]$.

If we let the random variable $N(t)$ be the number of removals (whether due to unit failure or scheduled replacement) in $(0, t]$, it is clear from Example 1 (of Subsection 9.2.1) that

$$P[N(t) = 0] = 1 - H(t) = [1 - F(t)][1 - G(t)]$$

and

$$P[N(t) = k] = H_k(t) - H_{k+1}(t), \qquad k = 1, 2, \ldots. \tag{9.28}$$

It also follows from the renewal theorem that

$$\lim_{t \to \infty} \frac{M(t)}{t} = \frac{1}{\int_0^\infty [1 - H(x)]dx} \tag{9.29}$$

where $M(t) = E[N(t)]$.

We denote the expected number of item removals because of failure in $(0, t]$ as $M_F(t)$ and the expected number of item removals because of planned replacement in $(0, t]$ as $M_P(t)$. It is then easy to show that

$$M_F(t) = pM(t) \quad \text{and} \quad M_P(t) = qM(t) \tag{9.30}$$

where $p = \int_0^\infty F(x)dG(x)$ and $q = 1 - p = \int_0^\infty G(x)dF(x)$ and hence

$$\lim_{t \to \infty} \frac{M_F(t)}{t} = \frac{\int_0^\infty F(x)dG(x)}{\int_0^\infty [1 - H(x)]dx}$$

and

$$\lim_{t \to \infty} \frac{M_P(t)}{t} = \frac{\int_0^\infty G(x)dF(x)}{\int_0^\infty [1 - H(x)]dx}. \tag{9.31}$$

It is of interest to find the distribution of T_1^F, the time when the first item failure occurs. Let $H_{T_1^F}(t) = P(T_1^F \leq t)$. Then $H_{T_1^F}(t)$ satisfies the integral equation

$$H_{T_1^F}(t) = \int_0^t [1 - G(\tau)]dF(\tau) + \int_0^t H_{T_1^F}(t - \tau)[1 - F(\tau)]dG(\tau), \tag{9.32}$$

where the first term on the right-hand side of (9.32) is the probability that the *first* replacement is due to a failure in $(0, t]$ and the second term gives the probability that the first item replacement at time τ is a planned replacement and that the first item failure occurs in the interval $(\tau, t]$, where $0 < \tau < t$.

If we define $b(s), c(s)$, and $h^*_{T_1^F}(s)$ as

$$b(s) = \int_0^\infty e^{-st}[1 - F(t)]dG(t)$$

$$c(s) = \int_0^\infty e^{-st}[1 - G(t)]dF(t)$$

and

$$h^*_{T_1^F}(s) = \int_0^\infty e^{-st}dH_{T_1^F}(t), \qquad (9.33)$$

and apply the Laplace-Stieltjes transform to both sides of (9.32), we obtain

$$h^*_{T_1^F}(s) = \frac{c(s)}{1 - b(s)}. \qquad (9.34)$$

For suitably chosen $F(t)$ and $G(t)$, it is possible to obtain an explicit formula for $H_{T_1^F}(t)$ by inverting $h^*_{T_1^F}(s)$ (see, e.g., Problems 1 and 3). If $p = \int_0^\infty F(t)dG(t)$ is sufficiently small and $F(t)$ and $G(t)$ have finite means, it can be shown that $P(T_1^F > t)$ is approximately equal to e^{-pt/μ_H}, where

$$\mu_H = \int_0^\infty tdH(t) = \int_0^\infty [1 - F(t)][1 - G(t)]dt.$$

Since $E(T_1^F) = \mu_H/p$, the approximation for $P(T_1^F > t)$ can be written as $e^{-t/E(T_1^F)}$. Results of this kind are a consequence of limit theorems for first passage times in regenerative processes (see, e.g., Chapter 6 of Gnedenko et al. [30]).

A quantity of interest is $R(\tau; t)$, the probability that the item will be *failure free without replacement* in the interval $[t, t + \tau]$ (this is also called *interval reliability*). It turns out that $R(\tau; t)$ satisfies the integral equation

$$R(\tau; t) = [1 - G(t)][1 - F(t + \tau)] + \int_0^t [1 - G(t - u)][1 - F(\tau + t - u)]dM(u),$$
$$(9.35)$$

where $M(t)$ satisfies the renewal equation

$$M(t) = \int_0^t [1 + M(t - x)]dH(x). \qquad (9.36)$$

The limiting interval reliability can be shown to be

$$\lim_{t \to \infty} R(\tau; t) = \frac{\int_0^\infty [1 - F(\tau + v)][1 - G(v)]dv}{\int_0^\infty [1 - F(v)][1 - G(v)]dv}. \qquad (9.37)$$

9.5 Example 4: Two-unit repairable system

In this section we reconsider the system of Example 2 of Subsection 8.1.2, namely, two-unit repairable system with exponentially distributed unit times to failure and repair.

It is readily verified that the following integral equations are satisfied by the $P_{ij}(t)$, $i, j = 0, 1, 2$:

$$P_{22}(t) = e^{-\lambda_2 t} + \int_0^t \lambda_2 e^{-\lambda_2 \tau} P_{12}(t - \tau) d\tau$$

$$P_{12}(t) = \int_0^t \mu_1 e^{-(\lambda_1+\mu_1)\tau} P_{22}(t - \tau) d\tau + \int_0^t \lambda_1 e^{-(\lambda_1+\mu_1)\tau} P_{02}(t - \tau) d\tau$$

$$P_{02}(t) = \int_0^t \mu_0 e^{-\mu_0 \tau} P_{12}(t - \tau) d\tau$$

$$P_{21}(t) = \int_0^t \lambda_2 e^{-\lambda_2 \tau} P_{11}(t - \tau) d\tau$$

$$P_{11}(t) = e^{-(\lambda_1+\mu_1)t} + \int_0^t \mu_1 e^{-(\lambda_1+\mu_1)\tau} P_{21}(t - \tau) d\tau$$

$$+ \int_0^t \lambda_1 e^{-(\lambda_1+\mu_1)\tau} P_{01}(t - \tau) d\tau$$

$$P_{01}(t) = \int_0^t \mu_0 e^{-\mu_0 \tau} P_{11}(t - \tau) d\tau$$

$$P_{20}(t) = \int_0^t \lambda_2 e^{-\lambda_2 \tau} P_{10}(t - \tau) d\tau$$

$$P_{10}(t) = \int_0^t \mu_1 e^{-(\lambda_1+\mu_1)\tau} P_{20}(t - \tau) d\tau + \int_0^t \lambda_1 e^{-(\lambda_1+\mu_1)\tau} P_{00}(t - \tau) d\tau$$

$$P_{00}(t) = e^{-\mu_0 t} + \int_0^t \mu_0 e^{-\mu_0 \tau} P_{10}(t - \tau) d\tau \ .$$

$$(9.38)$$

To find $P_{i2}(t)$, $i = 0, 1, 2$, we apply Laplace transforms to the first three equations of (9.38). We obtain

$$P_{22}^*(s) = \frac{1}{\lambda_2 + s} + \frac{\lambda_2}{\lambda_2 + s} P_{12}^*(s)$$

$$P_{12}^*(s) = \frac{\mu_1}{\lambda_1 + \mu_1 + s} P_{22}^*(s) + \frac{\lambda_1}{\lambda_1 + \mu_1 + s} P_{02}^*(s)$$

$$P_{02}^*(s) = \frac{\mu_0}{\mu_0 + s} P_{12}^*(s) \ . \qquad (9.39)$$

Solving these equations, one gets for each $P_{i2}^*(s)$ a ratio of two polynomials in

s, of degree 2 or less in the numerator and of degree 3 in the denominator. The next step is expanding into partial fractions as is done in Section 5.2. Similar sets of equations can be written down for $P_{i1}^*(s)$, $i = 0, 1, 2$ and $P_{i0}^*(s)$, $i = 0, 1, 2$.

By virtue of the assumption of exponentially distributed unit times to failure and repair, the stochastic process describing the state of the system over time is a Markov process. A Markov process is a fortiori a regenerative process, since each time point t during the development of the process is a point of regeneration. In setting up the integral Equations (9.38), we used those particular points of regeneration at which either a unit failure occurs or a unit repair is completed.

9.6 Example 5: One out of n repairable systems

Here we look back at Example 4 of Subsection 8.1.4 with the added assumption that $G_i(t)$ is the c.d.f. of the time to repair the ith unit.

Let $P_{i0}(t)$ be the probability that the system is in the good state E_0 at time t given that it just entered E_i, $i = 0, 1, 2, \ldots, n$ at time $t = 0$. It is readily verified that the $P_{i0}(t)$ satisfy the following set of integral equations:

$$P_{00}(t) = e^{-\Lambda t} + \sum_{i=1}^{n} \int_0^t \lambda_i e^{-\Lambda \tau} P_{i0}(t - \tau) d\tau$$

$$P_{i0}(t) = \int_0^t P_{00}(t - \tau) dG_i(\tau), \qquad i = 1, 2, \ldots, n. \qquad (9.40)$$

Applying Laplace transforms to these equations, we obtain

$$P_{00}^*(s) = \frac{1}{\Lambda + s}\left[1 + \sum_{i=1}^{n} \lambda_i P_{i0}^*(s)\right]$$

$$P_{i0}^*(s) = g_i^*(s) P_{00}^*(s), \qquad i = 1, 2, \ldots, n \qquad (9.41)$$

where $g_i^*(s) = \int_0^\infty e^{-st} dG_i(t)$ and

$$P_{i0}^*(s) = \int_0^\infty e^{-st} P_{i0}(t) dt, \qquad i = 0, 1, 2, \ldots, n.$$

It can be readily verified that

$$P_{00}^*(s) = \frac{1}{(\Lambda + s) - \sum_{i=1}^{n} \lambda_i g_i^*(s)}. \qquad (9.42)$$

It is possible, at least in principle, to invert $P_{00}^*(s)$ and thus obtain $P_{00}(t)$. In any case, we can readily compute $p_0 = \lim_{t \to \infty} P_{00}(t)$, the steady-state probability that the system is in the good state E_0. To compute p_0, let m_i equal the mean of the c.d.f. $G_i(t)$. Then, $g_i^*(s) = 1 - m_i s + o(s)$. Substituting in (9.42) and recalling that $\Lambda = \sum_{i=1}^n \lambda_i$, we obtain

$$P_{00}^*(s) = \frac{1}{(\Lambda + s) - \sum_{i=1}^n \lambda_i(1 - m_i s + o(s))} = \frac{1}{s\left[1 + \sum_{i=1}^n \lambda_i m_i\right] + o(s)}. \tag{9.43}$$

Hence,

$$p_0 = \lim_{t \to \infty} P_{00}(t) = \lim_{s \to 0} sP_{00}^*(s) = \frac{1}{1 + \sum_{i=1}^n \lambda_i m_i}. \tag{9.44}$$

This coincides with Equation (8.15).

9.7 Example 6: Section 7.3 revisited

The subject of this section is Example 2 of Subsection 8.2.2, which is also studied in Section 7.3.

Here, we consider anew the first passage time problem from E_2 into the failed state E_0 (where E_i, $i = 0, 1, 2$, is the state in which i units are good) for a two-unit repairable system with general unit repair time c.d.f. $G(t)$ and exponentially distributed unit time to failure. Let λ_i be the rate at which a change of state due to a unit failure occurs when the system is in state E_i, $i = 1, 2$. For the case of a two-unit system, with each unit active with common time to failure c.d.f. $F(t) = 1 - e^{-\lambda t}$, $t \geq 0$, $\lambda_2 = 2\lambda$ and $\lambda_1 = \lambda$. If one unit is active, the other is totally inactive and incapable of failure until called into service, $\lambda_2 = \lambda_1 = \lambda$. More generally, if a unit has failure rate λ when active and failure rate λ_s, when in standby status, then $\lambda_2 = \lambda + \lambda_s$ and $\lambda_1 = \lambda$.

Let $H_{i0}(t)$, $i = 1, 2$ be the c.d.f. of the first passage time into E_0 given that the system has just entered state E_i at time $t = 0$. We assert that

$$H_{20}(t) = \int_0^t H_{10}(t - \tau)\lambda_2 e^{-\lambda_2 \tau} d\tau$$

and

$$\tag{9.45}$$

$$H_{10}(t) = \int_0^t \lambda_1 e^{-\lambda_1 \tau}[1 - G(\tau)]d\tau + \int_0^t H_{20}(t - \tau)e^{-\lambda_1 \tau} dG(\tau).$$

In the first equation in (9.45), we start with the system in E_2. The system will pass into E_0 on or before time t if:

(i) The first unit failure occurs at some time τ, $0 < \tau < t$, the system thus passing into E_1.

(ii) The system passes into E_0 on or before time t, given that the system just entered E_1 at time τ.

Since $H_{10}(t - \tau)$ is the probability of (ii), integrating over all possible values of τ gives $\int_0^t H_{10}(t - \tau)\lambda_2 e^{-\lambda_2 \tau} d\tau$.

In the second equation in (9.45), we assume that the system has just entered E_1 at time $t = 0$. Clearly, the system will pass into E_0 on or before time t if:

(i) The failed unit is not repaired by time τ, $0 < \tau < t$, and the good unit fails at time τ. When this happens, the system goes from E_1 into E_0 in a single step. Integrating over all possible values of τ gives the first term on the right-hand side of the second equation in (9.45),

or

(ii) The good unit does not fail by time τ, $0 < \tau < t$, and repair of the failed unit takes place at time τ, and the system passes into E_0 on or before time t given that it entered E_2 at time τ. Integrating over all possible values of τ gives the second term on the right-hand side of the second equation in (9.45).

Applying Laplace-Stieltjes transforms to (9.45) and letting

$$h_{i0}^*(s) = \int_0^\infty e^{-st} dH_{i0}(t), \qquad i = 1, 2,$$

and

$$g^*(s) = \int_0^\infty e^{-st} dG(t),$$

it follows that

$$h_{20}^*(s) = \left(\frac{\lambda_2}{s + \lambda_2}\right) h_{10}^*(s) \tag{9.46}$$

$$h_{10}^*(s) = \frac{\lambda_1}{s + \lambda_1}[1 - g^*(s + \lambda_1)] + g^*(s + \lambda_1)h_{20}^*(s).$$

Therefore,

$$h_{20}^*(s) = \frac{\lambda_1 \lambda_2 [1 - g^*(s + \lambda_1)]}{(s + \lambda_1)[s + \lambda_2(1 - g^*(s + \lambda_1))]}. \tag{9.47}$$

This coincides with Equation (7.86), which was obtained previously by a more complicated analysis.

In many practical problems, $p = \int_0^\infty \lambda_1 e^{-\lambda_1 t}[1 - G(t)]dt$ (the probability of going from $E_1 \to E_0$ in a single step) is very small. If this is the case, $H_{20}(t) \approx 1 - e^{-\lambda_2 pt}$. For a proof of this result, see Chapter 6 of Gnedenko et al. [30]. We have previously noted this result in Section 7.3, Problem 4.

Now, consider the first passage time problem for the case where the system consists of two active dissimilar units with time to failure c.d.f.s $F_i(t) = 1 - e^{-\lambda_i t}$, $t \geq 0$, $i = 1, 2$, respectively, and associated unit time to repair c.d.f.s $G_i(t)$. Again, we assume that at time $t = 0$, both units are in a good state and are interested in the distribution of the length of time until both units are down simultaneously for the first time. Four states are needed to describe the system. These are:

$E_3 = (1,1)$ — both units are in a good state.

$E_2 = (1,0)$ — unit 1 is in a good state and unit 2 is in a failed state and under repair.

$E_1 = (0,1)$ — unit 1 is in a failed state and under repair and unit 2 is in a good state.

$E_0 = (0,0)$ — both units are in a failed state.

Let $H_{j0}(t)$ be the c.d.f. of the length of time to first enter E_0 given that the system has just entered E_j at time $t = 0$.

Analogously to the set of Equations (9.45), we find the following three integral equations are satisfied by the $H_{j0}(t)$, $j = 1, 2, 3$:

$$H_{30}(t) = \int_0^t \lambda_1 e^{-(\lambda_1+\lambda_2)\tau} H_{10}(t-\tau)d\tau + \int_0^t \lambda_2 e^{-(\lambda_1+\lambda_2)\tau} H_{20}(t-\tau)d\tau ,$$

$$H_{20}(t) = \int_0^t \lambda_1 e^{-\lambda_1\tau}[1-G_2(\tau)]d\tau + \int_0^t e^{-\lambda_1\tau} H_{30}(t-\tau)dG_2(\tau), \quad (9.48)$$

$$H_{10}(t) = \int_0^t \lambda_2 e^{-\lambda_2\tau}[1-G_1(\tau)]d\tau + \int_0^t e^{-\lambda_2\tau} H_{30}(t-\tau)dG_1(\tau).$$

Letting $h_{i0}^*(s) = \int_0^\infty e^{-st} dH_{i0}(t)$, $i = 1, 2, 3$ and $g_i^*(s) = \int_0^\infty e^{-st} dG_i(t)$, $i = 1, 2$, it follows that

$$h_{30}^*(s) = \frac{\lambda_1}{s+\lambda_1+\lambda_2} h_{10}^*(s) + \frac{\lambda_2}{s+\lambda_1+\lambda_2} h_{20}^*(s)$$

$$h_{20}^*(s) = \frac{\lambda_1}{s+\lambda_1}[1-g_2^*(s+\lambda_1)] + g_2^*(s+\lambda_1) h_{30}^*(s) \quad (9.49)$$

$$h_{10}^*(s) = \frac{\lambda_2}{s+\lambda_2}[1-g_1^*(s+\lambda_2)] + g_1^*(s+\lambda_2) h_{30}^*(s).$$

Solving for $h_{30}^*(s)$ gives

$$h_{30}^*(s) = \frac{\lambda_1\lambda_2 \left[\frac{1-g_1^*(s+\lambda_2)}{s+\lambda_2} + \frac{1-g_2^*(s+\lambda_1)}{s+\lambda_1}\right]}{s + \lambda_1[1-g_1^*(s+\lambda_2)] + \lambda_2[1-g_2^*(s+\lambda_1)]}. \quad (9.50)$$

The mean first passage time ℓ_{30}, from E_3 to E_0, is given by

$$\ell_{30} = -\left.\frac{d}{ds} h_{30}^*(s)\right]_{s=0} = \frac{1 + \frac{\lambda_1}{\lambda_2}[1-g_1^*(\lambda_2)] + \frac{\lambda_2}{\lambda_1}[1-g_2^*(\lambda_1)]}{\lambda_1[1-g_1^*(\lambda_2)] + \lambda_2[1-g_2^*(\lambda_1)]}. \quad (9.51)$$

For the particular case of exponential repair time with rates μ_i, $i = 1, 2$ one has $g_i^*(s) = \mu_i/(s + \mu_i)$ and so Equation (9.51) becomes

$$\ell_{30} = \frac{(\lambda_2 + \mu_1)(\lambda_1 + \mu_2) + \lambda_1(\lambda_1 + \mu_2) + \lambda_2(\lambda_2 + \mu_1)}{\lambda_1 \lambda_2 (\lambda_1 + \lambda_2 + \mu_1 + \mu_2)}. \tag{9.52}$$

Equations (9.51) and (9.52) can be obtained directly using the method of embedded Markov chains (see, e.g., Problems 3 and 5 of Section 8.2).

9.8 Example 7: First passage time distribution

The subject here is first passage time distribution from E_3 to E_0 for a three-unit repairable system, exponentially distributed unit times to failure and repair. For another treatment, see Section 7.2.

Let $H_{i0}(t)$, $i = 1, 2, 3$ be the c.d.f. of the time of first entrance into state E_0 (failed state) given that the system is in state E_i at time $t = 0$. The $H_{i0}(t)$ satisfy the following system of integral equations

$$H_{30}(t) = \int_0^t H_{20}(t - \tau)\lambda_3 e^{-\lambda_3 \tau} d\tau$$

$$H_{20}(t) = \int_0^t H_{30}(t - \tau)\mu_2 e^{-(\lambda_2 + \mu_2)\tau} d\tau$$

$$+ \int_0^t H_{10}(t - \tau)\lambda_2 e^{-(\lambda_2 + \mu_2)\tau} d\tau \tag{9.53}$$

$$H_{10}(t) = \int_0^t \lambda_1 e^{-(\lambda_1 + \mu_1)\tau} d\tau + \int_0^t H_{20}(t - \tau)\mu_1 e^{-(\lambda_1 + \mu_1)\tau} d\tau.$$

Applying Laplace-Stieltjes transforms to Equations (9.53) and letting $h_{i0}^*(s) = \int_0^\infty e^{-st} dH_{i0}(t)$, $i = 1, 2, 3$ yields

$$h_{30}^*(s) = \frac{\lambda_3}{\lambda_3 + s} h_{20}^*(s)$$

$$h_{20}^*(s) = \frac{\lambda_2}{\lambda_2 + \mu_2 + s} h_{10}^*(s) + \frac{\mu_2}{\lambda_2 + \mu_2 + s} h_{30}^*(s) \tag{9.54}$$

$$h_{10}^*(s) = \frac{\lambda_1}{\lambda_1 + \mu_1 + s} + \frac{\mu_1}{\lambda_1 + \mu_1 + s} h_{20}^*(s).$$

We leave it as a problem to the reader to find $h_{30}^*(s)$ and to then find $h_{30}(t)$, the p.d.f. of the first time to enter the failed state E_0 given that at $t = 0$ the system was in state E_3.

9.9 Problems and comments

Problems for Subsection 9.2.2

1. Show Equation (9.10) directly.

2. Suppose that the length of life of the items is exponential with rate λ. Show that $M(t) = \lambda t$,

   (a) Using Equation (9.7).

   (b) Using Equation (9.8) and then inverting $M^*(s)$.

3. Suppose that the length of life of the items is gamma distributed with parameters 2 and λ (i.e., the p.d.f. is $f(t) = \lambda^2 t e^{-\lambda t}$, $t \geq 0$).

   (a) Show that

$$M(t) = \frac{\lambda t}{2} - \frac{1}{4}(1 - e^{-2\lambda t})$$

   using Equation (9.8) and then inverting $M^*(s)$.

   (b) Show that

$$m(t) = \frac{\lambda}{2}(1 - e^{-2\lambda t})$$

   either by differentiating $M(t)$ or using $m(t) = \sum_{k=1}^{\infty} f_k(t)$.

4. Suppose that the length of the life of the items is a mixture of two independent exponential random variables (i.e., the p.d.f. is $f(t) = \alpha \lambda_1 e^{-\lambda_1 t} + (1 - \alpha)\lambda_2 e^{-\lambda_2 t}$, $t \geq 0$, $0 \leq \alpha \leq 1$). Find $M(t)$ and $m(t)$.

5. Equation (9.2) becomes $P_k(t) = \int_0^t \lambda e^{-\lambda \tau} P_{k-1}(t - \tau)d\tau$ when F is exponential. Show that the differential equation $P_k'(t) = -\lambda P_k(t) + \lambda P_{k-1}(t)$, $k = 1, 2, \ldots$ is obtained by differentiating this integral equation.

6. Verify, using Formula (9.5), that for exponential F, $P_k^*(s) = \lambda^k/(\lambda+s)^{k+1}$ and hence $P_k(t) = (\lambda t)^k e^{-\lambda t}/k!$.

7. Verify that $F_k(t)$, the c.d.f. of T_k (the time of the kth renewal) satisfies the integral equation $F_k(t) = \int_0^t F_{k-1}(t - \tau)dF(\tau)$, $k = 2, 3, \ldots$ and prove that $f_k^*(s)$, the Laplace-Stieltjes transform of $F_k(t)$, equals $[f^*(s)]^k$.

8. (Continuation of Problem 7). Invert $f_k^*(s)$ to find $f_k(t)$ for the exponential case and verify that $f_k(t) = \lambda(\lambda t)^{k-1}e^{-\lambda t}/(k - 1)!$, $k = 1, 2, \ldots, t \geq 0$.

9. Invert $f_k^*(s)$ to find $f_k(t)$ when $f(t) = \lambda(\lambda t)^{a-1}e^{-\lambda t}/\Gamma(a)$, $t \geq 0$, $a > 0$. Verify that

$$f_k(t) = \frac{\lambda(\lambda t)^{ka-1}e^{-\lambda t}}{\Gamma(ka)}, \qquad t \geq 0.$$

10. Invert $f_k^*(s)$ to find $f_k(t)$ when

$$f(t) = 1/A, \quad 0 \leq t \leq A$$
$$= 0, \quad \text{elsewhere.}$$

Verify that for $0 \le t \le kA$,

$$f_k(t) = \frac{1}{A^k(k-1)!}\left[\sum_{j=0}^{k}(-1)^j\binom{k}{j}(t-jA)_+^{k-1}\right]$$

where

$$(x)_+ = 0, \qquad x < 0$$
$$= x, \qquad x \ge 0.$$

11. Suppose that the life length of the item placed on test at $t = 0$ is distributed with c.d.f. $G(t)$ and p.d.f. $g(t)$, and that all replacement items have length of life distributed with c.d.f. $F(t)$ as before. The point process generated by successive failures is clearly a renewal process beginning with the first failure time T_1. However, since T_1 is distributed with c.d.f. $G(t)$ and all subsequent times between failures are distributed with c.d.f. $F(t)$, we call the stochastic process for $t \ge 0$ a *modified renewal process* (as contrasted with the *ordinary renewal process* for which $G(t) = F(t)$). Let $\tilde{M}(t)$ be the expected number of renewals in $(0, t]$ for the modified renewal process.

(a) Show that $\tilde{M}(t)$ satisfies the integral equation

$$\tilde{M}(t) = G(t) + \int_0^t F(t-\tau)d\tilde{M}(\tau).$$

(b) Show that $\tilde{M}(t)$ is related to $M(t)$, the expected number of renewals in $(0, t]$ for the ordinary renewal process, by the integral equation

$$\tilde{M}(t) = \int_0^t [1 + M(t-\tau)]dG(\tau).$$

(c) Show that
$$\tilde{M}^*(s) = g^*(s)/[s(1 - f^*(s))].$$

(d) Show that if $g(t) = [1 - F(t)]/\mu$, then $\tilde{M}^*(s) = 1/(\mu s^2)$ and hence $\tilde{M}(t) = t/\mu$ (and $\tilde{m}(t) = \tilde{M}'(t) = 1/\mu$) for all $t \ge 0$.

Comment. A modified renewal process for which the time to first renewal is distributed with p.d.f. $g(t) = [1 - F(t)]/\mu$ is called an *equilibrium renewal process*. The appropriateness of this name will become clear further on. It is interesting that for the equilibrium process $\tilde{M}_e(t) = t/\mu$, whereas for the ordinary renewal process we could only say that $\lim_{t\to\infty} M(t)/t = 1/\mu$ (the only ordinary renewal process with $M(t) = t/\mu$ is the one for which the time between renewals is exponential with mean μ). It is intuitively clear that $M(t)$ and $\tilde{M}_e(t)$ must be "close" as $t \to \infty$, since the equilibrium renewal process differs from the ordinary renewal process only in that the lifetime of the item placed in service at time $t = 0$ has c.d.f. $G(t) = \int_0^t [1 -$

$F(x)]dx/\mu$ instead of $F(t)$, but the lifetimes of subsequent replacements (beginning with the first one) are all assumed to have c.d.f. $F(t)$. As a matter of fact, if the c.d.f. $F(t)$ has mean μ and variance σ^2, $\lim_{t \to \infty}\{M(t) - \tilde{M}_e(t)\} = (\sigma^2 - \mu^2)/(2\mu^2)$.

12. Let $\tilde{N}(t)$ be the number of renewals in a modified renewal process. Let $\tilde{P}_k(t) = P[\tilde{N}(t) = k]$.

  (a) Show that $\tilde{P}_0(t) = 1 - G(t)$ and $\tilde{P}_k(t) = H_k(t) - H_{k+1}(t)$, $k = 1, 2, \ldots$
    where $H_k(t) = \int_0^t F_{k-1}(t - \tau)dG(\tau)$, and $F_0(t) \equiv 1$ for $t \geq 0$.

  (b) Show that $\tilde{M}(t) = \sum_{k=1}^{\infty} H_k(t)$.
    (It should be noted that $H_1(t) = G(t)$.)

13. Let $f_k^{(e)}(t)$ be the p.d.f. of the time to the kth renewal in the equilibrium renewal process. Show that

$$f_k^{(e)}(t) = \frac{F_{k-1}(t) - F_k(t)}{\mu} = \frac{P_{k-1}(t)}{\mu}, \quad k = 1, 2, \ldots,$$

where $F_0(t) \equiv 1$ for $t \geq 0$.

14. Suppose that the mean and standard deviation of the time between renewals in a certain renewal process is $\mu = 1$ day and $\sigma = 2$ days, respectively. What is the probability that the number of renewals in 100 days exceeds 140? Lies between 80 and 140? Is less than or equal to 50?

15. Suppose that a device is subject to shocks, where the times between shocks are independent random variables with common c.d.f. $F(t)$, which possesses finite mean μ and finite variance σ^2. Suppose that the conditional probability that the device fails (does not fail) given that a shock occurs is $p(q = 1 - p)$. It is assumed that the device has no memory, i.e., if it has survived $k - 1$ shocks, then the conditional probability of surviving the kth shock remains equal to q. Let T be the random variable for the length of time until the device fails.

  (a) Show that $G_T(t)$, the c.d.f. of T, is given by

$$G_T(t) = \sum_{k=1}^{\infty} \alpha_k F_k(t), \quad \text{where} \quad \alpha_k = q^{k-1}p, \quad k = 1, 2, \ldots.$$

  (b) Show that $g_T^*(s) = \int_0^{\infty} e^{-st}dG_T(t)$, the Laplace-Stieltjes transform of $G_T(t)$, is given by

$$g_T^*(s) = \frac{pf^*(s)}{1 - qf^*(s)}.$$

  (c) Define the normalized random variable $T' = pT/\mu$. Show that the c.d.f. of T' is given by $G_{T'}(t) = G(\mu t/p)$ and the the Laplace-Stieltjes transform of $G_{T'}(t)$ is given by

$$g_{T'}^*(s) = \frac{pf^*(ps/\mu)}{1 - qf^*(ps/\mu)}.$$

(d) Show that $\lim_{p\to 0} g_{T'}^*(s) = 1/(1+s)$ and hence that $G_{T'}(t)$ converges to the standardized exponential c.d.f. $1 - e^{-t}$, $t \geq 0$ as $p \to 0$ (equivalently, that $\lim_{p\to 0} P(pT/\mu \leq t) = 1 - e^{-t}$).

(e) Suppose that $\mu = 100$ hours and $p = .01$, what is the approximate value of the probability that the device survives 500 hours?

Comment. In (d), a variant of a fundamental result in probability theory known as Lévy's convergence theorem is used.

Problems for Subsection 9.2.3

1. Verify that $R(\tau; t) = e^{-\lambda\tau}$ when $f(t) = \lambda e^{-\lambda t}$, $t \geq 0$.
2. Verify that $R(\tau; t) = e^{-\lambda\tau}\left[1 + \frac{\lambda\tau}{2}\left(1 + e^{-2\lambda t}\right)\right]$, when $f(t) = \lambda^2 t e^{-\lambda t}$, $t \geq 0$.
3. Show that

$$\lim_{t\to\infty} R(\tau; t) = \int_\tau^\infty \frac{1 - F(x)}{\mu} \, dx.$$

4. Give details leading to Equations (9.16) through (9.21).
5. Find the p.d.f.s of A_t, B_t, and C_t when $f(t) = \lambda e^{-\lambda t}$, $t \geq 0$.
6. Find the limiting p.d.f. of C_t as $t \to \infty$ when f is gamma(a, λ), that is

$$f(t) = \frac{\lambda(\lambda t)^{a-1}e^{-\lambda t}}{\Gamma(a)}, \qquad t \geq 0, \quad a > 0.$$

7. Consider a lifetime distribution with mean μ and variance σ^2. Suppose that items whose length of life follow this distribution are placed on test and that failed items are replaced immediately by new items. After a "long" time T has elapsed, show that the mean age and mean remaining life of items on test at time T is approximately $[1 + (\sigma/\mu)^2]\mu/2$ and that the mean life of items on test at time T is approximately $[1 + (\sigma/\mu)^2]\mu$.

8. (Continuation of Problem 7.) Show for a renewal process with $f(t) = 1 - e^{-t/\theta}$, $t \geq 0$, that the mean age and mean remaining life of items on test at time T is approximately θ and that the mean life of items on test at time T is approximately 2θ for large T.

9. Suppose that we start observing an ordinary renewal process after it has been in progress for a "very long" time T. What is the justification for saying that the renewal process started with the item on test at time T is an equilibrium renewal process for all $t > T$?

10. Let $A_t^{(e)}$, $B_t^{(e)}$, and $C_t^{(e)}$ be, respectively, the age, remaining life, and total life length of the item on test at time t for the equilibrium renewal process. Prove the following six statements:

(a) $P(A_t^{(e)} \leq x) = \int_{t-x}^{t}\{1 - F(t-u)\}dM_e(u)$,   if $x < t$
    $\qquad\qquad\quad = 1$,   if $x \geq t$.

(b) $P(A_t^{(e)} \leq x) = \frac{1}{\mu}\int_0^x\{1 - F(w)\}dw$,   if $x < t$.

(c) $P(B_t^{(e)} \leq y) = F_e(t+y) - F_e(t) + \int_0^t\{F(t+y-u) - F(t-u)\}dM_e(u)$.

(d) $P(B_t^{(e)} \leq y) = \frac{1}{\mu}\int_0^y\{1 - F(v)\}dv$, $y > 0$.

(e) $P(C_t^{(e)} \leq z) = \int_0^z\{M_e(t) - M_e(t-x)\}dF(x)$,   if $z < t$.

(f) $P(C_t^{(e)} \leq z) = \frac{1}{\mu}\int_0^z x\,dF(x)$,   if $z < t$.

Comment. The only fact needed in (b), (d) and (f) is that $M_e(t) = t/\mu$ for all $t > 0$. According to (d), if the remaining life of the item placed on test at time $t = 0$ has p.d.f. $[1 - F(y)]/\mu$, then the remaining life of the item on test for any $t > 0$ has the same p.d.f. This is why we call a modified renewal process with $g_e(t) = [1 - F(t)]/\mu$ an equilibrium (or stationary) renewal process.

11. A system is composed of n components possessing common time to failure p.d.f. $f(t) = 1/(2A)$, $0 \leq t \leq 2A$. Failure of any component causes the system to fail and component failures are assumed to occur independently of each other. Assume that whenever a component fails, it is replaced instantaneously by a new component.

(a) Do the successive system failure times $T_1 < T_2 < T_3 < \cdots$ form a renewal process?

(b) Suppose that the procedure of replacing components as they fail has gone on for a long time T. Show that the probability that the system does not fail (i.e., that no component fails) in the interval $(T, T+u]$, $u > 0$, u/A small is given by $[1 - u/(2A)]^{2n}$ as $T \to \infty$.

(c) Verify, if u/A is "small" and n is "large," that the probability in (b) is approximately $e^{-nu/A}$.

(d) If $A = 1,000$, $n = 100$, what is the probability that the interval $(T, T+u]$ is failure free for $u = 10, 20, 30$ as $T \to \infty$?

Comment. The sequence of failure times generated by a particular component and its replacements is a renewal process. The sequence of system failure times arising from the failure of any of the components and their replacements is a stochastic process generated by the superposition of n component renewal processes. In this problem, we see that after a long time has elapsed (so that the system is in a "steady" state), the time to the next system failure and indeed the times between successive system failures is approximately exponentially distributed.

More generally, if a system consists of a large number n of components each having mean time to failure $\mu_1, \mu_2, \ldots, \mu_n$, then it can be shown under fairly modest conditions usually met in practice, that system failure times occur according to a Poisson process with rate $\lambda = \sum_{i=1}^{n} 1/\mu_i$ when the system has attained a steady (equilibrium) state. This has already been pointed out in Section 2.3. In this connection, it is interesting to read Example (a) on pp. 370–371 of Feller [27].

Problems for Section 9.3

1. Find $P_{11}(t)$ for the one-unit repairable system with exponentially distributed times to failure and repair.

2. Suppose that the time to failure p.d.f. is $f(t) = \lambda e^{-\lambda t}$, $t \geq 0$ and that the time to repair p.d.f. is $g(t) = \mu^2 t e^{-\mu t}$, $t \geq 0$. Find $P_{11}(t)$ for $\lambda = 5$, $\mu = 1$.

3. Suppose that the time to failure p.d.f. is $f(t) = \lambda^2 t e^{-\lambda t}$, $t \geq 0$ and that the time to repair p.d.f. is $g(t) = \mu e^{-\mu t}$, $t \geq 0$. Find $P_{11}(t)$ for $\lambda = 1$, $\mu = 4$.

4. Suppose that $F(t)$ and $G(t)$ have finite means $m_X = \int_0^\infty t\,dF(t)$ and $m_Y = \int_0^\infty t\,dG(t)$ and finite variances. Show that $\lim_{s \to 0} s P_{11}^*(s) = \lim_{t \to \infty} P_{11}(t)$ $= m_X/(m_X + m_Y)$.

5. Suppose that the system has just entered E_1 at time $t = 0$.

   (a) Show that the sequence of regeneration points at which system failure occurs is a modified renewal process, where the time to the first renewal (failure) is distributed with c.d.f. $F(t)$ and the times between subsequent renewals (failures) are independent random variables with common c.d.f. $H(t) = \int_0^t G(t - \tau)dF(\tau)$.

   (b) Show that the sequence of regeneration points at which system repair occurs is an ordinary renewal process, where the times between renewals (repairs) are independent random variables with common c.d.f. $H(t)$.

6. Let $N_{ij}(t)$, $i, j = 0, 1$ be the number of "visits" to state E_j in $(0, t]$ given that the one-unit repairable system has just entered E_i at time $t = 0$.

   (a) Show that

   $$P[N_{11}(t) = 0] = P[N_{00}(t) = 0] = 1 - H(t)$$
   $$P[N_{11}(t) = k] = P[N_{00}(t) = k] = H_k(t) - H_{k+1}(t), \quad k = 1, 2, \ldots,$$

   where $H_{k+1}(t) = \int_0^t H_k(t - \tau)dH(\tau)$, $\quad k = 1, 2, \ldots$

   (b) Show that

   $$P[N_{10}(t) = 0] = 1 - F(t)$$
   $$P[N_{10}(t) = k] = L_{F,k}(t) - L_{F,k+1}(t), \quad k = 1, 2, \ldots,$$

   where $L_{F,k}(t) = \int_0^t H_{k-1}(t - \tau)dF(\tau)$ and $H_0(t) \equiv 1$ for $t \geq 0$.

(c) Show that

$$P[N_{01}(t) = 0] = 1 - G(t)$$
$$P[N_{01}(t) = k] = L_{G,k}(t) - L_{G,k+1}(t), \quad k = 1, 2, \ldots,$$

where $L_{G,k}(t) = \int_0^t H_{k-1}(t - \tau)dG(\tau)$ and $H_0(t) \equiv 1$ for $t \geq 0$.

7. Let X_i be the length of time that the system is in the good state E_1, the ith time. Let Y_i be the length of time that the system is in the bad state E_0, the ith time. Show by a direct probabilistic argument that

(a) $P_{11}(t) = P(X_1 > t) + P(X_1 + Y_1 \leq t, \; X_1 + Y_1 + X_2 > t)$

$\quad + P(X_1 + Y_1 + X_2 + Y_2 \leq t, \; X_1 + Y_1 + X_2 + Y_2 + X_3 > t) + \cdots$

$\quad\quad = \; [1 - F(t)] + H(t) * [1 - F(t)] + H_2(t) * [1 - F(t)]$
$\quad\quad + \cdots + H_k(t) * [1 - F(t)] + \cdots$

where $H_k(t) * [1 - F(t)] = \int_0^t [1 - F(t - \tau)]dH_k(\tau)$.

(b) Taking the Laplace transform of both sides of (a), verify that $P_{11}^*(s)$ is as given in Equation (9.25).

(c) Find $P_{01}^*(s)$, $P_{00}^*(s)$, and $P_{10}^*(s)$ in a similar way.

8. Let $M_{ij}(t) = E[N_{ij}(t)]$.

(a) Verify that the following integral equations are satisfied by the $M_{ij}(t)$:

$$M_{11}(t) \;=\; \int_0^t M_{01}(t - x)dF(x)$$

$$M_{01}(t) \;=\; \int_0^t \{1 + M_{11}(t - x)\}dG(x)$$

$$M_{10}(t) \;=\; \int_0^t \{1 + M_{00}(t - x)\}dF(x)$$

$$M_{00}(t) \;=\; \int_0^t M_{10}(t - x)dG(x)\,.$$

(b) Express $m_{ij}^*(s) = \int_0^\infty e^{-st}dM_{ij}(t)$, the Laplace-Stieltjes transforms of $M_{ij}(t)$, in terms of $f^*(s)$ and $g^*(s)$.

9. Find $M_{11}(t)$ for a one-unit repairable system with $F(t) = 1 - e^{-\lambda t}$, $t \geq 0$ and $G(t) = 1 - e^{-\mu t}$, $t \geq 0$.

10. Find $M_{11}(t)$ for Problems 2 and 3.

11. Prove that

$$P_{10}(t) \;=\; M_{10}(t) - M_{11}(t)$$
$$P_{01}(t) \;=\; M_{01}(t) - M_{00}(t)\,.$$

12. (a) Show that $M_{11}(t) = \sum_{k=1}^\infty H_k(t)$.

(b) Verify that $M_{11}(t)$ satisfies the renewal equation

$$M_{11}(t) = \int_0^t [1 + M_{11}(t - \tau)]dH(\tau).$$

(c) Find $m_{11}^*(s)$ by taking Laplace-Stieltjes transforms of the equations for $M_{11}(t)$ in (a) and (b) and verify that this is the same as found in Problem 8(b).

(d) Find $M_{01}(t)$, $M_{10}(t)$, and $M_{00}(t)$ in a similar way.

13. Let $R_{ij}(\tau;t)$ be the probability that the system will be in state E_j throughout the time interval $(t, t+\tau]$, given that the system is initially in E_i. Show that

$$R_{11}(\tau;t) = 1 - F(t+\tau) + \int_0^t [1 - F(t+\tau-u)]dM_{11}(u)$$

$$R_{01}(\tau;t) = \int_0^t [1 - F(t+\tau-u)]dM_{01}(u)$$

$$R_{10}(\tau;t) = \int_0^t [1 - G(t+\tau-u)]dM_{10}(u)$$

$$R_{00}(\tau;t) = 1 - G(t+\tau) + \int_0^t [1 - G(t+\tau-u)]dM_{00}(u).$$

Also show that

$$\lim_{t\to\infty} R_{11}(\tau;t) = \lim_{t\to\infty} R_{01}(\tau;t) = \frac{1}{m_X + m_Y}\int_\tau^\infty [1 - F(w)]dw$$

and

$$\lim_{t\to\infty} R_{10}(\tau;t) = \lim_{t\to\infty} R_{00}(\tau;t) = \frac{1}{m_X + m_Y}\int_\tau^\infty [1 - G(w)]dw.$$

Comment. The quantities $R_{11}(\tau;t)$ and $R_{01}(\tau;t)$ are, respectively, the interval reliabilities given that the system is initially in E_1 or E_0. Likewise $S_{10}(\tau;t) = 1 - R_{10}(\tau;t)$ is the probability that the system is in the good state E_1 at time t or, if in the bad state E_0 at time t, is repaired at some time during the interval $(t, t+\tau]$, given that the system is initially in state E_1.

14. Given a one-unit repairable system with time to failure p.d.f. $f(t) = \lambda e^{-\lambda t}$, $t \geq 0$ and with time to repair p.d.f. $g(t) = \mu e^{-\mu t}$, $t \geq 0$. Find $R_{11}(\tau;t)$ and $\lim_{t\to\infty} R_{11}(\tau;t)$.
Hint: Use the results in Problem 9.

15. Given a one-unit repairable system with time to failure p.d.f. $f(t) = \lambda^2 t e^{-\lambda t}$, $t \geq 0$ and time to repair p.d.f. $g(t) = \mu^2 t e^{-\mu t}$, $t \geq 0$. Find $\lim_{t\to\infty} R_{11}(\tau;t)$ and $\lim_{t\to\infty} S_{10}(\tau;t)$.

16. In Example 2 and in Problems 1–15, it is assumed that repair is initiated

instantaneously upon the occurrence of system failure. Suppose, instead, that a waiting time W distributed with c.d.f. $W(t)$ elapses before repair is initiated. Let the time to failure and time to repair c.d.f.s be $F(t)$ and $G(t)$, respectively. Let E_2, E_1, E_0 be, respectively, states in which the system is good, system is failed but waiting to be repaired, system is failed and under repair. Let $P_{ij}(t)$, $i, j = 0, 1, 2$ be the probability that the system is in state E_j at time t given that it has just entered E_i at time $t = 0$.

(a) Show that the $P_{ij}(t)$ satisfy the integral equations:

$$P_{22}(t) = [1 - F(t)] + \int_0^t P_{12}(t - \tau)dF(\tau)$$

$$P_{12}(t) = \int_0^t P_{02}(t - \tau)dW(\tau)$$

$$P_{02}(t) = \int_0^t P_{22}(t - \tau)dG(\tau)$$

$$P_{21}(t) = \int_0^t P_{11}(t - \tau)dF(\tau)$$

$$P_{11}(t) = [1 - W(t)] + \int_0^t P_{01}(t - \tau)dW(\tau)$$

$$P_{01}(t) = \int_0^t P_{21}(t - \tau)dG(\tau)$$

$$P_{20}(t) = \int_0^t P_{10}(t - \tau)dF(\tau)$$

$$P_{10}(t) = \int_0^t P_{00}(t - \tau)dW(\tau)$$

$$P_{00}(t) = [1 - G(t)] + \int_0^t P_{20}(t - \tau)dG(\tau).$$

(b) Express $P_{ij}^*(s)$, the Laplace transform of $P_{ij}(t)$, in terms of $f^*(s), g^*(s)$, and $w^*(s)$, the Laplace-Stieltjes transform of $F(t), G(t)$, and $W(t)$, respectively.

17. Suppose that $W(t)$ has finite mean $m_W = \int_0^\infty t dW(t)$. Show that

$$\lim_{s \to 0} s P_{12}^*(s) = \lim_{t \to \infty} P_{12}(t) = \frac{m_X}{m_X + m_W + m_Y}.$$

Problems for Section 9.4

1. Suppose that $F(t) = 1 - e^{-\lambda t}$, $t \geq 0$ and $G(t)$ is general. Show that $h_{T_1^F}^*(s) = \lambda/(\lambda + s)$, and hence that the p.d.f. $h_{T_1^F}(t) = \lambda e^{-\lambda t}$, $t \geq 0$. Why is this result obvious from basic properties of the Poisson process?

2. Suppose that $F(t)$ is general and that $G(t) = 1 - e^{-\mu t}$, $t \geq 0$. Show that

$$h^*_{T^F_1}(s) = \frac{(s + \mu)f^*(s + \mu)}{s + \mu f^*(s + \mu)}.$$

3. (Continuation of Problem 2.) Suppose that $f(t) = F'(t) = \lambda^2 t e^{-\lambda t}$, $t \geq 0$.

(a) Show that

$$h^*_{T^F_1}(s) = \frac{\lambda^2}{s^2 + (2\lambda + \mu)s + \lambda^2},$$

and hence that

$$h_{T^F_1}(t) = \frac{\lambda^2}{s_1 - s_2}\left\{e^{s_1 t} - e^{s_2 t}\right\},$$

where

$$s_1 = \frac{-(2\lambda + \mu) + \sqrt{4\lambda\mu + \mu^2}}{2}$$

and

$$s_2 = \frac{-(2\lambda + \mu) - \sqrt{4\lambda\mu + \mu^2}}{2}.$$

(b) Show that $E(T^F_1)$ and $\mathrm{Var}(T^F_1)$, where T^F_1 is the time to the first replacement due to failure are, respectively,

$$E(T^F_1) = (2\lambda + \mu)/\lambda^2 \quad \text{and} \quad \mathrm{Var}(T^F_1) = (2\lambda^2 + 4\lambda\mu + \mu^2)/\lambda^4.$$

(c) Show that, if μ/λ is "large," T^F_1 is approximately exponentially distributed with mean $(2\lambda + \mu)/\lambda^2$.

4. Using Formula (9.34) for $h^*_{T^F_1}(s)$, show that

$$E(T^F_1) = -\frac{d}{ds}\left[h^*_{T^F_1}(s)\right]_{s=0} = \frac{\int_0^\infty [1 - F(t)][1 - G(t)]dt}{\int_0^\infty F(t)dG(t)}.$$

5. Suppose that $f(t) = F'(t) = \lambda^2 t e^{-\lambda t}$, $t \geq 0$, with $\lambda = .01$ and $g(t) = \mu e^{-\mu t}$, $t \geq 0$, with $\mu = .09$.

(a) Using Problem 3, find the exact probability that the first unscheduled failure occurs within the first 100 hours.

(b) Compute an approximate value for this probability using the formula

$$P(T^F_1 > t) \approx e^{-t/E(T^F_1)} = e^{-\lambda^2 t/(2\lambda + \mu)}.$$

6. Suppose that $f(t) = F'(t) = \lambda^{k+1} t^k e^{-\lambda t}/k!$, $t \geq 0$ and $G(t) = 1 - e^{-\mu t}$, $t \geq 0$. Show that $E(T^F_1) = [(1 + x)^{k+1} - 1]/\mu$, where $x = \mu/\lambda$.

7. Let $N_F(t)$ be the number of replacements due to item failure in $(0, t]$. Let $P^F_k(t) = P[N_F(t) = k]$, $k = 0, 1, 2, \ldots$.

(a) Show that the $P_k^F(t)$ satisfy the following system of integral equations:

$$P_0^F(t) = [1 - F(t)][1 - G(t)] + \int_0^t P_0^F(t - \tau)[1 - F(\tau)]dG(\tau)$$

$$P_k^F(t) = \int_0^t P_{k-1}^F(t - \tau)[1 - G(\tau)]dF(\tau)$$

$$+ \int_0^t P_k^F(t - \tau)[1 - F(\tau)]dG(\tau), \quad k = 1, 2, \ldots.$$

(b) In addition to the transforms $b(s)$ and $c(s)$, defined in Equations (9.33), define the transforms $A(s)$ and $P_k^{F^*}(s)$ as:

$$A(s) = \int_0^\infty e^{-st}[1 - F(t)][1 - G(t)]dt$$

$$P_k^{F^*}(s) = \int_0^\infty e^{-st} P_k^F(t)dt.$$

(c) Verify the identity $-sA(s) + b(s) + c(s) = 1$.

(d) Applying Laplace transforms to both sides of the equation in (a), verify that

$$P_0^{F^*}(s) = \frac{A(s)}{1 - b(s)}$$

and

$$P_k^{F^*}(s) = \frac{A(s)}{1 - b(s)} \left[\frac{c(s)}{1 - b(s)} \right]^k.$$

8. (a) Show that

$$P(T_1^F > t) = 1 - H(t) + B(t) * [1 - H(t)]$$
$$+ B_2(t) * [1 - H(t)] + \cdots + B_k(t) * [1 - H(t)] + \cdots,$$

where

$$H(t) = 1 - [1 - F(t)][1 - G(t)],$$

$$B(t) = \int_0^t [1 - F(\tau)]dG(\tau),$$

$$B_k(t) = \int_0^t B_{k-1}(t - \tau)dB(\tau), \quad k = 2, 3, \ldots,$$

and

$$B_k(t) * [1 - H(t)] = \int_0^t [1 - H(t - \tau)]dB_k(\tau).$$

(b) Take Laplace transforms of both sides of the equation in (a) and thus obtain the expression for $P_0^{F^*}(s)$ obtained in Problem 7(d).

(c) Obtain Formula (9.34) for $h_{T_1^F}^*(s)$ from (b) and Problem 7(c).

9. (a) Let $M_F(t) = E[N_F(t)]$ be the expected number of removals in $(0, t]$ because of item failure. Show that $M_F^*(s) = \int_0^\infty e^{-st} M_F(t)dt$ satisfies the equation $M_F^*(s) = c(s)/(s^2 A(s))$.

   (b) Show that (a) implies that

$$\lim_{t \to \infty} \frac{M_F(t)}{t} = \frac{c(0)}{A(0)} = \frac{\int_0^\infty F(x)dG(x)}{\int_0^\infty [1 - F(x)][1 - G(x)]dx} .$$

10. Let $T_1^F, T_2^F, \ldots, T_k^F, \ldots$ be the times of occurrence of the first item failure, second item failure, etc. Let $H_1^F(t), H_2^F(t), \ldots, H_k^F(t), \ldots$ be the corresponding c.d.f.s of $T_1^F, T_2^F, \ldots, T_k^F, \ldots$.

   (a) Show that

$$P_0^F(t) = 1 - H_1^F(t)$$

   and

$$P_k^F(t) = H_k^F(t) - H_{k+1}^F(t), \qquad k = 1, 2, \ldots.$$

   By taking Laplace transforms of both sides, obtain an alternative derivation of 7(d).

   (b) Also obtain 9(a) from the relationship

$$M_F(t) = \sum_{k=1}^{\infty} H_k^F(t).$$

11. Verify Equation (9.35).

12. Suppose that $f(t) = F'(t) = \lambda^2 t e^{-\lambda t}$, $t \geq 0$ and $G(t) = 1 - e^{-\mu t}$, $t \geq 0$. Show that

$$\lim_{t \to \infty} R(\tau; t) = \frac{\lambda^2 e^{-\lambda \tau}[1 + \lambda \tau(\lambda + \mu)]}{(2\lambda + \mu)(\lambda + \mu)^2}.$$

13. Consider a system for which the time to failure is distributed with c.d.f. $F(t)$, if it is not maintained. The time between maintenance actions is distributed with c.d.f. $G(t)$. If a scheduled preventive maintenance action occurs prior to system failure, the length of time to bring the system back to new is a random variable with c.d.f. $V_1(t)$ and mean m_1. If, however, a system failure occurs prior to a scheduled preventive maintenance action, the length of time to bring the system back to new is a random variable with c.d.f. $V_0(t)$ and mean m_0. Let E_2 be the state in which the system is operating satisfactorily and not being repaired or maintained. Let E_1 be the state in which the system is operative and undergoing preventive maintenance to make it as good as new. Let E_0 be the state in which the system has failed and is being repaired. Let $P_{ij}(t)$, $i = 0, 1, 2$, be the probability that the system is in state E_j at time t, given that it just entered E_i at time $t = 0$.

(a) Show that the $P_{ij}(t)$ satisfy the following system of integral equations:

$$P_{22}(t) = [1 - F(t)][1 - G(t)] + \int_0^t P_{12}(t - \tau)[1 - F(\tau)]dG(\tau)$$

$$+ \int_0^t P_{02}(t - \tau)[1 - G(\tau)]dF(\tau).$$

$$P_{12}(t) = \int_0^t P_{22}(t - \tau)dV_1(\tau)$$

$$P_{02}(t) = \int_0^t P_{22}(t - \tau)dV_0(\tau).$$

$$P_{21}(t) = \int_0^t P_{11}(t - \tau)[1 - F(\tau)]dG(\tau)$$

$$+ \int_0^t P_{01}(t - \tau)[1 - G(\tau)]dF(\tau)$$

$$P_{11}(t) = [1 - V_1(t)] + \int_0^t P_{21}(t - \tau)dV_1(\tau)$$

$$P_{01}(t) = \int_0^t P_{21}(t - \tau)dV_0(\tau).$$

$$P_{20}(t) = \int_0^t P_{10}(t - \tau)[1 - F(\tau)]dG(\tau)$$

$$+ \int_0^t P_{00}(t - \tau)[1 - G(\tau)]dF(\tau)$$

$$P_{10}(t) = \int_0^t P_{20}(t - \tau)dV_1(\tau)$$

$$P_{00}(t) = [1 - V_0(t)] + \int_0^t P_{20}(t - \tau)dV_0(\tau).$$

(b) Taking Laplace transforms in (a), show that

$$P_{22}^*(s) = \frac{A(s)}{1 - b(s)v_1^*(s) - c(s)v_0^*(s)},$$

where

$$P_{22}^*(s) = \int_0^\infty e^{-st}P_{22}(t)dt,$$

$A(s), b(s)$, and $c(s)$ are as previously defined, and

$$v_i^*(s) = \int_0^\infty e^{-st}dV_i(t), \qquad i = 0, 1.$$

(c) Show that

$$\lim_{s \to 0} s P_{22}^*(s) = \lim_{t \to \infty} P_{22}(t) =$$

$$\frac{\int_0^\infty [1-F(t)][1-G(t)]dt}{\int_0^\infty [1-F(t)][1-G(t)]dt + m_1 \int_0^\infty G(t)dF(t) + m_0 \int_0^\infty F(t)dG(t)}.$$

Problems for Section 9.5

1. Verify the set of Equations (9.38) relating the $P_{ij}(t)$.
2. Find $P_{22}^*(s)$ and then invert the transform to find $P_{22}(t)$.
3. Verify that the transforms $P_{ij}^*(s)$ satisfy the matrix equation:
   $\mathbb{P}^*(s) = A(s)\,\mathbb{P}^*(s) + B(s)$, where

$$\mathbb{P}^*(s) = \begin{pmatrix} P_{00}^*(s) & P_{01}^*(s) & P_{02}^*(s) \\ P_{10}^*(s) & P_{11}^*(s) & P_{12}^*(s) \\ P_{20}^*(s) & P_{21}^*(s) & P_{22}^*(s) \end{pmatrix}$$

$$A(s) = \begin{pmatrix} 0 & \frac{\mu_0}{\mu_0+s} & 0 \\ \frac{\lambda_1}{\lambda_1+\mu_1+s} & 0 & \frac{\mu_1}{\lambda_1+\mu_1+s} \\ 0 & \frac{\lambda_2}{\lambda_2+s} & 0 \end{pmatrix}$$

$$B(s) = \begin{pmatrix} \frac{1}{\mu_0+s} & 0 & 0 \\ 0 & \frac{1}{\lambda_1+\mu_1+s} & 0 \\ 0 & 0 & \frac{1}{\lambda_2+s} \end{pmatrix}.$$

4. Consider the n-unit repairable system with exponentially distributed unit times to failure and repair (see Table 6.3). Verify that the following integral

equations are satisfied by $P_{ij}(t)$, $i, j = 0, 1, 2, \ldots, n$.

$$P_{nn}(t) = e^{-\lambda_n t} + \int_0^t \lambda_n e^{-\lambda_n \tau} P_{n-1,n}(t-\tau)d\tau$$

$$P_{nj}(t) = \int_0^t \lambda_n e^{-\lambda_n \tau} P_{n-1,j}(t-\tau)d\tau, \qquad j = 0, 1, 2, \ldots, n-1$$

$$P_{00}(t) = e^{-\mu_0 t} + \int_0^t \mu_0 e^{-\mu_0 \tau} P_{10}(t-\tau)d\tau$$

$$P_{0j}(t) = \int_0^t \mu_0 e^{-\mu_0 \tau} P_{1j}(t-\tau)d\tau, \qquad j = 1, 2, \ldots, n$$

$$P_{ii}(t) = e^{-(\lambda_i + \mu_i)t} + \int_0^t \mu_i e^{-(\lambda_i + \mu_i)\tau} P_{i+1,i}(t-\tau)d\tau$$

$$+ \int_0^t \lambda_i e^{-(\lambda_i + \mu_i)\tau} P_{i-1,i}(t-\tau)d\tau, \qquad i = 1, 2, \ldots, n-1$$

$$P_{ij}(t) = \int_0^t \mu_i e^{-(\lambda_i + \mu_i)\tau} P_{i+1,j}(t-\tau)d\tau$$

$$+ \int_0^t \lambda_i e^{-(\lambda_i + \mu_i)\tau} P_{i-1,j}(t-\tau)d\tau, \qquad i = 1, 2, \ldots, n-1;$$

$$j \neq i, \quad j = 0, 1, 2, \ldots, n.$$

Problems for Section 9.6

1. Verify that the integral equations satisfied by $P_{0j}(t)$ and $P_{ij}(t)$ (where $P_{ij}(t)$ is the probability that the system is in state E_j at time t given that the system has just entered state E_i at time $t = 0$) are:

$$P_{0j}(t) = \sum_{i=1}^n \int_0^t \lambda_i e^{-\Lambda \tau} P_{ij}(t-\tau)d\tau$$

$$P_{ij}(t) = \int_0^t P_{0j}(t-\tau)dG_i(\tau), \qquad i \neq j$$

$$P_{jj}(t) = [1 - G_j(t)] + \int_0^t P_{0j}(t-\tau)dG_j(\tau),$$

where $i = 1, 2, \ldots, n$ and $j = 0, 1, 2, \ldots, n$.

2. (Continuation of Problem 1.) Verify that

$$P_{0j}^*(s) = \frac{\lambda_j[1 - g_j^*(s)]}{s[\Lambda + s - \sum_{i=1}^n \lambda_i g_i^*(s)]} = \frac{\lambda_j[1 - g_j^*(s)]}{s} P_{00}^*(s).$$

3. Prove that $p_j = \lim_{t \to \infty} P_{ij}(t) = \lambda_j m_j / [1 + \sum_{i=1}^n \lambda_i m_i]$, $j = 1, 2, \ldots, n$.

4. Compute $P_{00}(t)$ for the special case of a system composed of $n = 2$ compo-

nents, and where $F_i(t) = 1 - e^{-\lambda_i t}$, $t \geq 0$, $i = 1, 2$, and $G_i(t) = 1 - e^{-\mu_i t}$, $t \geq 0$, $i = 1, 2$.

5. Obtain Equation (9.44) directly without using Equations (9.40).

Problems for Section 9.7

1. Suppose for simplicity that the repair time p.d.f. $g(t) = G'(t)$ exists.

   (a) Show that $h_{20}(t)$, the p.d.f. of the first passage time from $E_2 \to E_0$, can be expressed as:

   $$h_{20}(t) = u(t) + v_1(t) * u(t) + v_2(t) * u(t) + \cdots + v_k(t) * u(t) + \cdots,$$

   where

   $$u(t) = \int_0^t \lambda_2 e^{-\lambda_2 \tau}[1 - G(t - \tau)]\lambda_1 e^{-\lambda_1(t-\tau)}d\tau = a(t) * b(t),$$

   with $a(t) = \lambda_2 e^{-\lambda_2 t}$ and $b(t) = [1 - G(t)]\lambda_1 e^{-\lambda_1 t}$, $t \geq 0$; where

   $$v_1(t) = \int_0^t \lambda_2 e^{-\lambda_2 \tau} e^{-\lambda_1(t-\tau)}g(t - \tau)d\tau = a(t) * c(t),$$

   with $c(t) = e^{-\lambda_1 t}g(t)$, $t \geq 0$; and where

   $$v_k(t) = \int_0^t v_{k-1}(t - \tau)v_1(\tau)d\tau, \quad k = 2, 3, \ldots.$$

   Hint: $u(t)\Delta t$ is the probability that the first passage time from $E_2 \to E_0$ occurs in $(t, t + \Delta t)$ via the route $E_2 \to E_1 \to E_0$; $v(t) * u(t)\Delta t$ is the probability that the first passage time from $E_2 \to E_0$ occurs in $(t, t+\Delta t)$ via the route $E_2 \to E_1 \to E_2 \to E_1 \to E_0$, etc.

   (b) Take Laplace transforms of both sides of the equation in (a) and thus rederive Equation (9.47).

2. Consider a two-unit system, where both units are active and each unit possesses common time to failure p.d.f. $f(t) = \lambda e^{-\lambda t}$, $t \geq 0$. Suppose that there is a single repairman with unit time to repair p.d.f. $g(t) = \mu e^{-\mu t}$, $t \geq 0$.

   (a) Show that $h_{20}(t)$, the p.d.f. of the first passage time from $E_2 \to E_0$, can be expressed as:

   $$h_{20}(t) = \sum_{k=1}^{\infty} \alpha_k c_k(t),$$

   where

   $$\alpha_k = \left(\frac{\lambda}{\lambda + \mu}\right)\left(\frac{\mu}{\lambda + \mu}\right)^{k-1}$$

and where $c_1(t)$ is the convolution of the p.d.f.s $2\lambda e^{-2\lambda t}$ and $(\lambda+\mu)e^{-(\lambda+\mu)t}$, i.e.,

$$c_1(t) = \int_0^t 2\lambda e^{-2\lambda \tau}(\lambda + \mu)e^{-(\lambda+\mu)(t-\tau)}d\tau$$

and $c_k(t)$ is the k-fold convolution of $c_1(t)$.

(b) Assign a probabilistic meaning to each term in the expansion of $h_{20}(t)$.

3. Consider a two-unit redundant system, with one unit active with time to failure p.d.f. $f(t) = \lambda e^{-\lambda t}$, $t \geq 0$, and one unit inactive until called into service. The inactive unit, when it becomes active, also has the time to failure p.d.f. $f(t)$. Suppose that upon failure of the active unit, a switch is used to activate the inactive unit and that the switch has probability p of failing to function (hence causing system failure) and probability $q = 1 - p$ of functioning properly. Suppose that the repair time p.d.f. is given by $g(t)$. Find $H_{20}(t)$, the c.d.f. of the first passage time from state E_2 (both units good) into state E_0 (system is down).
Show that $h_{20}^*(s) = \int_0^\infty e^{-st}dH_{20}(t)$ is given by the equation

$$h_{20}^*(s) = \frac{\lambda[p(s+\lambda) + q\lambda(1 - g^*(s+\lambda))]}{(s+\lambda)[s+\lambda(1 - qg^*(s+\lambda))]} .$$

4. (a) Evaluate $h_{20}^*(s)$ for the special case where $G(t) = 1 - e^{-\mu t}$, $t \geq 0$ and find $H_{20}(t)$.

(b) For $p = 0$, the result here coincides with the result in Problem 9.4.3. Explain why.

5. Compute the mean and variance of ξ_{20}, the first passage time from E_2 into E_0, for the special cases where $G(t) = 1 - e^{-\mu t}$, $t \geq 0$ and

$$\begin{aligned} G(t) &= 0, & t < t_0 \\ &= 1, & t \geq t_0. \end{aligned}$$

6. Using the set of Equations (9.48), find the p.d.f. $h_{30}(t)$, if $G_i(t) = 1 - e^{-\mu_i t}$, $t \geq 0$, $i = 1, 2$.

7. Consider a two-unit repairable system, with both units active, each with exponentially distributed time to failure density function $f(t) = \lambda e^{-\lambda t}$, $t \geq 0$. Suppose that the system continues to function satisfactorily if one of the units fails and that unit failures are not instantaneously detected because inspections are not performed continuously but rather at random times. More precisely, the times between successive inspections are assumed to be independent random variables drawn from the density function $a(t) = \nu e^{-\nu t}$, $t \geq 0$. The act of inspection is assumed to take no time at all. The time to repair a failed unit is assumed to be distributed with general c.d.f. $G(t)$. Let E_3 be the state in which both units are good; E_2, the state in which one unit is good, one unit is down but not under repair; E_1, the state in which one unit is good, one unit is down and under repair; E_0, the state in which both

units are down (system failure). Let $H_{j0}(t)$, $j = 1, 2, 3$, be the c.d.f. of the first time to enter E_0 given that E_j has just been entered into at $t = 0$.

(a) Show that $H_{30}(t)$, $H_{20}(t)$, $H_{10}(t)$ satisfy the three integral equations:

$$H_{30}(t) = \int_0^t 2\lambda e^{-2\lambda\tau} H_{20}(t - \tau)d\tau$$

$$H_{20}(t) = \frac{\lambda}{\lambda + \nu}\left[1 - e^{-(\lambda+\nu)t}\right] + \int_0^t \nu e^{-(\lambda+\nu)\tau} H_{10}(t - \tau)d\tau$$

$$H_{10}(t) = \int_0^t \lambda e^{-\lambda\tau}[1 - G(\tau)]d\tau + \int_0^t e^{-\lambda\tau} H_{30}(t - \tau)dG(\tau).$$

(b) Solve for $h_{j0}^*(s) = \int_0^\infty e^{-st}dH_{j0}(t)$.

(c) Find an explicit solution for $H_{30}(t)$ if $G(t) = 1 - e^{-\mu t}$, $t \geq 0$.

(d) Find ℓ_{30}, the mean first passage time from E_3 into E_0. Use the method of embedded Markov chains.

8. Consider the same two-unit system as in Problem 7, with the following changes:

(i) There is a fixed time T between inspections, i.e., inspections are at T, $2T$, $3T$, ...;

(ii) The time to repair a failed unit is $g(t) = \mu e^{-\mu t}$, $t \geq 0$ with $\mu > \lambda$.

(a) Using the same definition of states, show that

$$H_{30}(t) = \left(1 - e^{-\lambda t}\right)^2, \qquad 0 \leq t \leq T$$

$$H_{10}(t) = \int_0^t \mu e^{-(\lambda+\mu)\tau} h_{30}(t - \tau)d\tau + \int_0^t e^{-(\lambda+\mu)\tau}\lambda d\tau$$

$$= \left(1 - e^{-\lambda t}\right)^2 + \frac{\lambda}{\mu - \lambda}\left(e^{-2\lambda t} - e^{-(\lambda+\mu)t}\right), \quad 0 \leq t \leq T$$

$$H_{30}(t) = H_{30}(T) + P_{33}(T)H_{30}(t-T) + P_{31}(T)H_{10}(t-T), \quad t > T$$

$$H_{10}(t) = H_{10}(T) + P_{13}(T)H_{30}(t-T) + P_{11}(T)H_{10}(t-T), \quad t > T$$

where $P_{ij}(T)$, $i, j = 1, 3$, is the probability that the system is in E_j at time T given that it was in E_i at time 0.

(b) Show that

$$P_{33}(T) = e^{-2\lambda T}$$

$$P_{31}(T) = 2(1 - e^{-\lambda T})e^{-\lambda T}$$

$$P_{13}(T) = \frac{\mu}{\mu - \lambda}\left[e^{-2\lambda T} - e^{-(\lambda+\mu)T}\right]$$

$$P_{11}(T) = \left(\frac{\mu + \lambda}{\mu - \lambda}\right)e^{-(\lambda+\mu)T} + 2e^{-\lambda T} - 2\left(\frac{\mu}{\mu - \lambda}\right)e^{-2\lambda T}.$$

(c) Suppose that $P_{ij}(kT)$, $i, j = 1, 3$ is the probability that the system is in E_j at time kT given that it was in E_i at time 0. Show that

$$
\begin{aligned}
P_{33}(kT) &= P_{33}(T)P_{33}[(k-1)T] + P_{31}(T)P_{13}[(k-1)T] \\
P_{31}(kT) &= P_{33}(T)P_{31}[(k-1)T] + P_{31}(T)P_{11}[(k-1)T] \\
P_{13}(kT) &= P_{11}(T)P_{13}[(k-1)T] + P_{13}(T)P_{33}[(k-1)T] \\
P_{11}(kT) &= P_{11}(T)P_{11}[(k-1)T] + P_{13}(T)P_{31}[(k-1)T], \\
& \quad k = 2, 3, \ldots.
\end{aligned}
$$

(d) Show that for $kT \le t \le (k+1)T$,

$$
\begin{aligned}
H_{30}(t) = H_{30}(kT) \; &+ \; P_{33}(kT)H_{30}(t - kT) \\
&+ \; P_{31}(kT)H_{10}(t - kT)
\end{aligned}
$$

and

$$
\begin{aligned}
H_{10}(t) = H_{10}(kT) \; &+ \; P_{13}(kT)H_{30}(t - kT) \\
&+ \; P_{11}(kT)H_{10}(t - kT).
\end{aligned}
$$

(e) Show that

$$
H_{30}(kT) = 1 - P_{33}(kT) - P_{31}(kT)
$$

and

$$
H_{10}(kT) = 1 - P_{13}(kT) - P_{11}(kT).
$$

9. At time $t = 0$, a unit A, whose lifetime is distributed with c.d.f. $F(t)$, is placed in service. An identical unit B, which is inactive (i.e., does not age until called into service) is in reserve. When A fails, the hitherto inactive reserve unit B, whose length of life when in service is also distributed with c.d.f. $F(t)$, is placed in service (perfect switching is assumed) and repair is begun at once on A. Time to repair a unit is distributed with c.d.f. $G(t)$. Repair is assumed to restore a failed unit to its original state, i.e., to make it as good as new. If B fails before repair on A is completed, we say that system failure has occurred. If, however, repair of A is completed before B fails, A becomes the inactive reserve. When B does fail, the current reserve unit A is placed into service and repair is begun at once on B. If A fails before repair on B is completed, we say that system failure has occurred. If, however, repair of B is completed before A fails, B becomes the inactive reserve, which is called into service when A fails. This process in which A and B alternate as active and (inactive) reserve units goes on until system failure occurs, i.e., at the moment when the unit currently in service fails before repair is completed on the currently failed unit.

(a) Let $\Phi_{20}(t)$ be the probability that system failure occurs on or before time t, given that we start with two new units, one active, the other an inactive reserve, at $t = 0$. Let $\Phi_{10}(t)$ be the probability that system failure occurs on or before time t, given that the active unit has just failed at $t = 0$, and that the inactive unit has just been activated and that repair on the failed unit has just been initiated. Show that $\Phi_{20}(t)$ and $\Phi_{10}(t)$ satisfy the pair of integral equations:

$$\Phi_{20}(t) = \int_0^t \Phi_{10}(t - \tau)dF(\tau),$$

$$\Phi_{10}(t) = \int_0^t [1 - G(\tau)]dF(\tau) + \int_0^t \Phi_{10}(t - \tau)G(\tau)dF(\tau).$$

(b) Define the transforms

$$\phi_{20}^*(s) = \int_0^\infty e^{-st}d\Phi_{20}(t)$$

$$\phi_{10}^*(s) = \int_0^\infty e^{-st}d\Phi_{10}(t)$$

$$f^*(s) = \int_0^\infty e^{-st}dF(t)$$

$$a(s) = \int_0^\infty e^{-st}G(t)dF(t).$$

Show that

$$\phi_{10}^*(s) = \frac{f^*(s) - a(s)}{1 - a(s)}$$

and that

$$\phi_{20}^*(s) = f^*(s) \cdot \frac{f^*(s) - a(s)}{1 - a(s)}.$$

(c) Let T_{20} be the time until system failure occurs. Show that $E(T_{20}) = (p + 1)\mu_F/p$, where $\mu_F = \int_0^\infty tdF(t)$ and $p = \int_0^\infty F(t)dG(t)$.

Comment. It can be shown that if p is small, then $P(T_{20} > t) \approx e^{-pt/\mu_F}$.

Problems for Section 9.8

1. Consider the n-unit repairable system (with exponentially distributed unit times to failure and repair) discussed in Chapter 6. Let $H_{i0}(t)$, $i = 1, 2, \ldots, n$ be the c.d.f. of the time to first enter the failed state E_0 given that the system is in state E_i at time $t = 0$. Let

$$h_{i0}^*(s) = \int_0^\infty e^{-st}dH_{i0}(t).$$

Show that

$$h_{no}^*(s) = \frac{\lambda_n}{\lambda_n + s} h_{n-1,0}^*(s),$$

$$h_{io}^*(s) = \frac{\lambda_i}{\lambda_i + \mu_i + s} h_{i-1,0}^*(s) + \frac{\mu_i}{\lambda_i + \mu_i + s} h_{i+1,0}^*(s)$$

$$i = 2, \ldots, n-1,$$

$$h_{10}^*(s) = \frac{\lambda_1}{\lambda_1 + \mu_1 + s} + \frac{\mu_1}{\lambda_1 + \mu_1 + s} h_{20}^*(s).$$

2. Consider a semi-Markov process with associated embedded Markov chain consisting of states $\{i\}$, $i = 0, 1, 2, \ldots, n$, transition probabilities matrix $\mathbf{P} = (p_{ij})$, and matrix of transition time c.d.f.s $\mathbb{F}(t) = (F_{ij}(t))$, $i, j = 0, 1, 2, \ldots, n$.
   Let $m_{ij} = \int_0^\infty t \, dF_{ij}(t)$ and $m_i = \sum_{j=0}^n p_{ij} m_{ij}$. Let $H_{ij}(t)$ be the c.d.f. of the length of time to first enter E_j given that the system is initially in E_i. If $j = i$, $H_{ii}(t)$ is the recurrent time c.d.f. between successive entries into E_i. Verify that

$$H_{ij}(t) = p_{ij} F_{ij}(t) + \sum_{k \neq j} \int_0^t H_{kj}(t - \tau) p_{ik} dF_{ik}(\tau).$$

3. Let $h_{ij}^*(s) = \int_0^\infty e^{-st} dH_{ij}(t)$ and $f_{ij}^*(s) = \int_0^\infty e^{-st} dF_{ij}(t)$. Verify that

$$h_{ij}^*(s) = p_{ij} f_{ij}^*(s) + \sum_{k \neq j} p_{ik} f_{ik}^*(s) h_{kj}^*(s).$$

4. Let the random variable ξ_{ij} be the time to pass into E_j for the first time, given that the system is initially in E_i. Let $\ell_{ij} = E(\xi_{ij})$ and $\ell_{ij}^{(2)} = E(\xi_{ij}^2)$. Verify by differentiating both sides of the equation in Problem 3 with respect to s and setting $s = 0$ that

$$\ell_{ij} = m_i + \sum_{k \neq j} p_{ik} \ell_{kj}$$

and

$$\ell_{ij}^{(2)} = \sum_{k \neq j} p_{ik} \left(\ell_{kj}^{(2)} + 2 m_{ik} \ell_{kj} \right) + m_i^{(2)},$$

where $m_i^{(2)} = \sum_{k=0}^n p_{ik} m_{ik}^{(2)}$ and where $m_{ik}^{(2)} = \int_0^\infty t^2 dF_{ik}(t)$.

5. Show that

$$\ell_{jj} = \frac{1}{\pi_j} \sum_{k=0}^n \pi_k m_k$$

and

$$\ell_{jj}^{(2)} = \frac{1}{\pi_j} \left\{ \sum_{k=0}^n \pi_k m_k^{(2)} + 2 \sum_{k \neq i} \sum_i \pi_i p_{ik} m_{ik} \ell_{kj} \right\},$$

where the $\{\pi_i\}$ are the stationary probabilities for the embedded Markov chain.

6. A special case of the stochastic process considered in Problems 2, 3, 4 and 5 is a discrete Markov chain consisting of $n+1$ states $\{0, 1, 2, \ldots, n\}$ with transition probabilities matrix $\mathbf{P} = (p_{ij})$. What is the interpretation of ℓ_{ij}, $\ell_{ij}^{(2)}$, ℓ_{jj}, and $\ell_{jj}^{(2)}$ for the Markov chain? Find formulae for these quantities by making appropriate substitutions in the formulae in Problems 4 and 5.

7. Let $N_{ij}(t)$ be the number of times that the system considered in Problem 2 enters state E_j in the time interval $(0, t]$ given that the system is initially in state E_i. Let $M_{ij}(t) = E[N_{ij}(t)]$. Prove that

$$M_{ij}(t) = \int_0^t \{1 + M_{jj}(t - \tau)\} p_{ij} dF_{ij}(\tau)$$

$$+ \int_0^t \sum_{k \ne j} M_{kj}(t - \tau) p_{ik} dF_{ik}(\tau)$$

$$= p_{ij} F_{ij}(t) + \sum_{k=0}^n \int_0^t M_{kj}(t - \tau) p_{ik} dF_{ik}(\tau)$$

$$= p_{ij} F_{ij}(t) + \sum_{k=0}^n \int_0^t p_{ik} F_{ik}(t - \tau) dM_{kj}(\tau).$$

8. Show that Equation (9.10) is a special case of the formula in the preceding problem, when there is only one system state and that the equations in Problem 6 of Section 9.3 for the two-state system are also a special case of the formula in the preceding problem.

9. Prove that

$$M_{ij}(t) = \int_0^t \{1 + M_{jj}(t - \tau)\} dH_{ij}(\tau)$$

$$= H_{ij}(t) + \int_0^t H_{ij}(t - \tau) dM_{jj}(\tau).$$

References

[1] R.E. Barlow, L.C. Hunter, and F. Proschan. Optimum redundancy when components are subject to two kinds of failure. *Journal of the Society for Industrial and Applied Mathematics*, 11:64–73, 1963.

[2] R.E. Barlow and F. Proschan. *Statistical Theory of Reliability and Life Testing*. To Begin With, Silver Spring, MD, 1981.

[3] R.E. Barlow and F. Proschan. *Mathematical Theory of Reliability*. SIAM, Society for Industrial and Applied Mathematics, Philadelphia, 1996.

[4] J. Beirlant, Y. Goegebeur, J. Segers, and J. Teugels. *Statistics of Extremes: Theory and Applications*. John Wiley, New York, 2004.

[5] P.J. Bickel and K. A. Doksum. *Mathematical Statistics*. Pearson Prentice Hall, Upper Saddle River, New Jersey, 2nd edition, 2007.

[6] P. Billingsley. *Probability and Measure*. John Wiley, New York, 1995.

[7] Z.W. Birnbaum and S.C. Saunders. A statistical model for life-length of materials. *Journal of the American Statistical Association*, 53:151–160, 1958.

[8] E. Brockmeyer, H.L. Holstrom, and A. Jensen. *The Life and Works of A.K. Erlang*. Transactions of the Danish Academy of Technical Sciences, Copenhagen, 1948.

[9] D.R. Cox. *Renewal Theory*. Methuen and Co., London, 1970.

[10] D.R. Cox and H.D. Miller. *The Theory of Stochastic Processes*. Chapman and Hall/CRC, New York, 2001.

[11] D.R. Cox and W.L. Smith. On the superposition of renewal processes. *Biometrika*, 41:91–99, 1954.

[12] M.J. Crowder. *Classical Competing Risks*. John Wiley, New York, 2001.

[13] M.J. Crowder, A.C. Kimber, R.L. Smith, and T.J. Sweeting. *Statistical Analysis of Reliability Data*. Chapman and Hall/CRC, London, 2nd edition, 1994.

[14] H.A. David and M.L. Moeschberger. *Theory of Competing Risks*. Lu-

brecht and Cramer Ltd., Port Jervis, New York, 1978.

[15] L. de Haan and A. Ferreira. *Extreme Value Theory: An Introduction.* Springer, New York, 2006.

[16] R.F. Drenick. The failure law of complex equipment. *Journal of the Society for Industrial and Applied Mathematics*, 8:680–690, 1960.

[17] P.P.G. Dyke. *An Introduction to Laplace Transforms and Fourier Series.* Springer, New York, 2002.

[18] B. Epstein. Application of the theory of extreme values in fracture problems. *Journal of the American Statistical Association*, 43:403–412, 1948.

[19] B. Epstein. Statistical aspects of fracture problems. *Journal of Applied Physics*, 19:140–147, 1948.

[20] B. Epstein. The exponential distribution and its role in life testing. *Industrial Quality Control*, 15:4–9, 1958.

[21] B. Epstein. Elements of the theory of extreme values. *Technometrics*, 2:27–41, 1960.

[22] B. Epstein. Formulae for the mean time between failures and repairs of repairable redundant systems. In *Proc. of IEEE*, volume 53, pages 731–32, 1965.

[23] B. Epstein and H. Brooks. The theory of extreme values and its implications in the study of the dielectric strength of paper capacitors. *Journal of Applied Physics*, 19:544–550, 1948.

[24] B. Epstein and J. Hosford. Reliability of some two-unit redundant systems. In *Proc. of the Sixth National Symposium on Reliability and Quality Control*, pages 469–476, 1960.

[25] A.J. Fabens. The solution of queueing and inventory models by semi-Markov processes. *Journal of the Royal Statistical Society, Series B*, 23:113–127, 1961.

[26] W. Feller. *An Introduction to Probability Theory and Its Applications, Vol. I.* John Wiley, New York, 3rd edition, 1968.

[27] W. Feller. *An Introduction to Probability Theory and Its Applications, Vol. II.* John Wiley, New York, 2nd edition, 1971.

[28] D.P. Gaver. Time to failure and availability of paralleled systems with repair. *IEEE Transactions on Reliability*, R-12(2):30–38, 1963.

[29] I. Gertsbakh. *Reliability Theory with Applications to Preventive Maintenance.* Springer, New York, 2nd edition, 2006.

[30] B.V. Gnedenko, Y.K. Belayev, and A.D. Solovyev. *Mathematical Methods*

of Reliability Theory. Academic Press, New York, 1969.

[31] E.J. Gumbel. *Statistics of Extremes*. Columbia University Press, New York, 1958.

[32] W.S. Jewell. Markov renewal programming I: Formulation, finite return models. *Operations Research*, 11:938–948, 1963.

[33] W.S. Jewell. Markov renewal programming II: Infinite return models, example. *Operations Research*, 11:949–971, 1963.

[34] V.V. Kalashnikov. *Topics in Regenerative Processes*. CRC Press Inc., New York, 1994.

[35] S. Karlin and H.M. Taylor. *A First Course in Stochastic Processes*. Academic Press, New York, 2nd edition, 1975.

[36] D.G. Kendall. Some problems in the theory of queues. *Journal of the Royal Statistical Society, Series B*, 13:151–185, 1951.

[37] D.G. Kendall. Stochastic processes occurring in the theory of queues and their analysis by the method of the imbedded Markov chain. *Annals of Mathematical Statistics*, 24:338–354, 1953.

[38] J. Korevaar. *Tauberian Theory: A Century of Developments*. Springer, New York, 2004.

[39] E.L. Lehmann and G. Casella. *Theory of Point Estimation*. Springer-Verlag, New York, 2nd edition, 2003.

[40] P. Lévy. Systèmes semi-Markoviens à au plus une infinité denombrable d'états possibles. In *Proc. Int. Congr. Math., Amsterdam*, volume 3, pages 416–423, 1954.

[41] D. Perry, W. Stadje, and S. Zacks. First-exit times for increasing compound processes. *Stochastic Models*, 15:5:977–992, 1999.

[42] M. Pintilie. *Competing Risks: A Practical Perspective*. John Wiley, New York, 2006.

[43] N.U. Prabhu. *Stochastic Processes: Basic Theory and Its Applications*. World Scientific, Hackensack, New Jersey, 2007.

[44] R. Pyke. Markov renewal processes: definitions and preliminary properties. *Annals of Mathematical Statistics*, 32:1231–1242, 1961.

[45] R. Pyke. Markov renewal processes with finitely many states. *Annals of Mathematical Statistics*, 32:1243–1259, 1961.

[46] R.L. Smith and I. Weissman. *Extreme Values: Theory and Applications*. To appear in Chapman and Hall/CRC, London, 2008.

[47] W.L. Smith. Regenerative stochastic processes. In *Proc. Roy. Soc., Lon-*

don, Ser. A, volume 232, pages 6–31, 1955.

[48] D.W. Stroock. *Introduction to Markov Processes*. Springer, New York, 2005.

[49] L. Takács. Some investigations concerning recurrent stochastic processes of a certain type. *Magyar Tud. Akad. Kutato Int. Kozl*, 3:115–128, 1954.

[50] L. Takács. On a sojourn time problem in the theory of stochastic processes. *Transactions of the American Mathematical Society*, 93:531–540, 1959.

[51] L. Takács. *Introduction to the Theory of Queues*. Oxford University Press, New York, 1982.

[52] G.H. Weiss. On a semi-Markovian process with a particular application to reliability theory. Technical Report 4351, U.S. Naval Ordnance Laboratory, White Oak, Maryland, 1960.

[53] G.H. Weiss. A problem in equipment maintenance. *Management Science*, 8:266–277, 1962.

[54] G.H. Weiss. Optimal periodic inspection programs for randomly failing equipment. *Journal of Res. of the Nat. Bur. of Standards*, 67B:223–228, 1963.

[55] G.H. Weiss and M. Zelen. A stochastic model for the interpretation of clinical trials. Technical Report 385, U.S. Army Math. Res. Center, Madison, Wisconsin, 1963.

[56] S. Zacks. *Introduction to Reliability Analysis*. Springer-Verlag, New York, 1992.

Index

Milton Keynes UK
Ingram Content Group UK Ltd.
UKHW020313111024
449327UK00040B/905